RADICAL RITUAL

RADICAL
RITUAL

How Burning Man Changed the World

NEIL SHISTER

COUNTERPOINT

Berkeley, California

Radical Ritual

Library of Congress Cataloging-in-Publication Data
Names: Shister, Neil, author.
Title: Radical ritual : how burning man changed the world / Neil Shister.
Description: First hardcover edition. | Berkeley, California : Counterpoint,
2019. | Includes index.
Identifiers: LCCN 2018051326 | ISBN 9781640092198
Subjects: LCSH: Burning Man (Festival) | Burning Man (Festival)—
Interviews. | Art festivals—Social aspects—Nevada—Black Rock
Desert. | Art festivals—Political aspects—Nevada—Black Rock Desert. |
Shister, Neil.
Classification: LCC NX510.N48 S55 2019 | DDC 700.74—dc23
LC record available at https://lccn.loc.gov/2018051326

Jacket design by Sarah Brody
Book design by Jordan Koluch

COUNTERPOINT
2560 Ninth Street, Suite 318
Berkeley, CA 94710
www.counterpointpress.com

Printed in the United States of America
Distributed by Publishers Group West

10 9 8 7 6 5 4 3 2 1

To Cait, who gets me to all places, including Burning Man

Collaboration is the soul of culture, but the system divides people.

LARRY HARVEY

CONTENTS

I: In the Beginning

II: Chaos and Control:
The Burning Man Experience

III: Burning Man's Utopian Vision

I

IN THE BEGINNING

1

The Creation Myth

The hero comes back from this mysterious adventure with the power to bestow boons on his
fellow man.

JOSEPH CAMPBELL,
The Hero with a Thousand Faces

The essential thing to understand about the whole Burning Man
phenomenon is that nobody saw it coming.

Not even your looniest, acid-addled Haight-Ashbury whack
job in the wildest of hallucinations could have foreseen in 1986
the chain of events stemming from an impromptu bonfire on a
San Francisco beach. It would lead three decades later to official

Washington, where the august Smithsonian Institution devoted the entire Renwick Gallery—housed in America's oldest art museum, a few steps from the White House—to an exhibition that traced back to that bonfire.

The person who least anticipated the magnitude of what was starting was the man who began it all. Larry Harvey, with a close friend and their young sons, lit that first fire in part to heal his broken heart. It was an act of self-medication. From that unlikely spark, a full-blown culture has taken form. Hundreds of thousands of celebrants have partaken of it in the Nevada desert over the years, close to a million people if you count sister events all over the world. A full-fledged, multi-million-dollar apparatus exists to spread that culture. The annual celebration, unnamed in the beginning, is now a global brand. But, to repeat, nobody foresaw any of this.

Were we back in biblical days, this tale might mark the start of a religion. One can imagine, say, somewhere in the Sinai wilderness, a crew like Harvey and his tiny cadre of fellow Founders. A visionary prophet with a scruffy band of acolytes, evangelizing a message of salvation that mesmerized growing multitudes.

Indeed, Harvey et al. could be regarded as another installment in a long list of American nonconformists who rewrote the scriptures as they crossed the continent. Maybe the Puritan elders could keep their progeny toeing the line back in Boston, but the kids shucked off those theological shackles when they hit the frontier. Enthusiasm consistently trumped piety in the great trek westward. When they finally reached the Pacific, with nowhere farther to go, what started as radical nonconformity gradually went mainstream as it ricocheted back east. To that point, some might compare the Renwick exhibition—the symbolic moment when power cedes respectability to outlaw renegades from the outback—to when the Roman emperor Constantine gave the official seal of approval to Christianity.

That, however, would be a wrong reading of the intent of the Founders. If religion is about ancestral reverence and supernatural awe, then Burning Man is about human agency in the moment. It is postmodern rather than theological. In 1966, as Larry Harvey was coming of age, the cover of *Time* magazine wondered aloud, "Is God Dead?" To dare ask such a question, forget about the answer, called into doubt civilization's most venerated certainty. Paradigms that had held society intact— secular and theological—were losing their grip. The Founders were hip to this. But they weren't out to start a religion. Their intention wasn't to allay anxiety by invoking some new supernatural entity. Just the opposite. The whole raison d'être of Burning Man was to sow doubt.

In the Burning Man cosmology, all truth is relative and all narratives equally credible. The culture posits no eschatological destination. Assumptions are never to be trusted on authority, only tested through experience. The sole orthodoxy is that there is no deference owed orthodoxy. Destiny for good or ill is the product of human acts. Thus, and herein resides the supreme tenet, people are capable of creating . . . anything.

The origins of Burning Man were a result of a specific place and moment. San Francisco in the mid-1980s was poised on a historical cusp. The magical mystery tour of the counterculture Sixties was about to combine with the digital wizardry taking form down the peninsula in Silicon Valley. These two vectors slammed into each other like particles in a cyclotron, fusing a new element from soul and science. One primal expression of it would emerge as Burning Man.

The Bay Area—port of call to whalers and gold rushers, communist longshoremen and vanguard gay activists—breeds more eccentricity pound-for-pound than anywhere else in America. Personal re-invention is in the water. Call it liberation or license, social rules are pliant here. Maybe such broad-mindedness is a natural adaptation, the response to a geological landscape that can split open without warning. "Living on a fault line," as one local observer understood, "is no easy business." Toleration, a mere virtue elsewhere, is more a strategy for mutual survival. That, along with natural beauty, a temperate climate, and nearby vineyards, makes for an existential wonderland. "It is and has everything," wrote Dylan Thomas. "And all the people are open and friendly."

By 1986, as Harvey and his pal Jerry James were hammering together scrap wood for the first Burn, the last wave of the Sixties counterculture that crested in San Francisco had pretty much ebbed. The ardor of insurrection was spent. The rising generation, who grew up with Ronald Reagan, didn't want to overthrow the system, they wanted to own it. Still, among the elders who had known the glory days, vestiges of rebellion lingered.

Meanwhile, south of the city in towns like Mountain View and Cupertino, a new swell was building. The digital epoch was poised to kick into overdrive. In 1986, IBM introduced the original laptop (weighing twelve pounds, it was a commercial dud). Detroit's car industry, linchpin of America's industrial might, was aging out but didn't yet know it. "The days of slow-moving technology, stable markets, and mass production processes that once allowed our industries to thrive in a sheltered environment have long since passed," Harold Shapiro, president of the University of Michigan, warned auto execs in his home state that year. They weren't listening. Who wants to hear their days are numbered? The electrical engineers and computer scientists at Stanford, however, got it. As did the smart venture capital money along Sand Hill Road in Palo Alto that was laying down tech bets.

Larry Harvey, of late a San Francisco bike messenger and griller of hot dogs at a grunge club, was enduring a long, painful romantic break-up in 1986. His had been no garden-variety love affair. "I lost my soul," he would say to me as we sat some three decades later in his cramped, art-filled apartment. No longer was he the dark-haired, narrow-faced young man who shows up in early photographs, handsome in a gentle way with soft features fixed in an ethereal gaze. Approaching seventy now, grayish hair dangling over his forehead *à la* Julius Caesar bangs, he still retained that patina of vulnerability, but he had survived (and provoked) too many close calls to look angelic. The descriptor I'd use is "wizened." A not-so-closeted cynic who managed to still project an air of innocence.

I was there to interview him for this book. We would talk at length perhaps a half dozen times. These days he was much in demand, as a speaker, as an eminence, as a celebrity. The following day he was off somewhere overseas, to participate in a panel discussion with a couple of Nobel Prize winners. I was lucky to get this session; it had to be squeezed into his schedule while he packed for the trip and his housekeeper cleaned the kitchen.

The one-bedroom flat was modest, a vintage kind of place that needed refreshing. Books lay piled on tables. The smell of cigarettes hung heavy. His living room looked out on Alamo Square, famous for its much-photographed Victorians known as the Painted Ladies of Postcard Row. My hunch was that the flat was rented rather than owned; I could imagine developers drooling to get their hands on the building.

The tale he was telling me, the origins of that first Burn, was ancient and much revised history. As ever, he spoke in extended monologues and abrupt sound bites, a ragged tempo of conversational counterpoint that left listeners to intuit much of his meaning. Even so, I thought I detected a dollop of residual pain in the romantic narrative he told me that afternoon. Many a night, he recounted, he'd stare reverentially at his beloved as she slept. Just watching her. "My soul drained out of my eyes into her."

By way of biographical data, Harvey was born in 1948 of parents whose identities he never knew. Scant ancestry information came from a DNA analysis: lots of English, some black Irish, a little Spanish. One aspect of temperament that might have been inherited, since even as a child he longed to play on a big stage, was the ambition gene. "Larry needs to be famous," recalled a friend who knew him long before Burning Man, "and feel that he has moved society in some way."[1]

He grew up outside Portland, Oregon, one of two boys adopted by Author and Katherine Harvey, expatriate Nebraskans who had fled the Dust Bowl. His father, by trade a carpenter and mason, was a man of strong ethics. Stoic and honest. And emotionally distant. "Semi-literate and so fierce in his autonomy," Harvey once wrote, "so insistent on his integrity as someone self-made, that not through a mean heart, but perhaps because of a blind one, he neglected to make contact with those around him."[2] His mother was equally reserved. "No one would speak about deeper feelings, certainly not in relation to one another." A photo of his family arranged around the dinner table, Thanksgiving perhaps or Christmas, shows a ring of stern-faced adults straight out of *American Gothic* with, in the foreground, a preteen Larry shamelessly mugging for the camera.

His practical father was less than receptive to the schemes Larry continually hatched. Like when he wanted to build a maze in the cornfield and invite neighbors to "encounter each other in a wonderful way." One can imagine Author's puzzlement: to what end? Harvey's answer was that, since the people in the neighborhood were barely acquainted, they'd all get caught up in the adventure and get to know each other. In grade school, he'd stage plays of his own design. "I'd dream up scenery and recruit people who had talent for doing things." Even back then, Harvey was foreshadowing Burning Man.

"I think," Harvey said of his admittedly unhappy childhood, "the folks were afraid of me." In Brian Doherty's fine book *This Is Burning Man*, Harvey portrays himself as a displaced child. The family iconoclast. The house intellectual who read the *Encyclopedia Britannica* for fun. "He was smarter than his parents," Doherty concludes. "He knew it and he is pretty sure they knew it, too."[3]

The army drafted him after high school. Being intelligent, he was trained as a regimental clerk. On board the flight to Germany, his first posting, he had an LSD-induced epiphany. "I saw myself going to Vietnam and dying." He made up his mind not to let this happen. The challenge was to figure a way to get out of the service without giving himself away. From his desk at headquarters processing paper, he watched others fail because

their ploys were too obvious. So, displaying the tactical shrewd-
ness for which he would later become famous, Harvey deviously
turned the system inside out. He blundered, screwing up assign-
ments not through carelessness but from overwrought zeal. His
super-sloppy salutes were executed with excessive enthusiasm.
He wore a Tyrolean hat in inappropriate situations, explaining
that he was carrying out orders to befriend the host Germans.
His desire to please was excruciating, his humiliation from failure so sincere he couldn't be suspected of insubordination. He
convinced his superiors that his gaffes were rooted in determination to be a good solider. "I realized that we were their tools
and, since it's a poor carpenter who would blame his hammer if
he smashed his thumb, how could they blame somebody who is
so zealous?" The strategy worked.

After months of this stuff, and sending him to the camp
shrink for anti-anxiety meds (he swallowed an overdose by design one day to fall asleep at his desk), the captain brought Harvey aside to deliver bad news. "I feel awful because I know how
hard you tried," he told him. "Once in a while, though, there's
somebody who just doesn't fit. What I'm saying, son, is that it's
time for you to leave." Harvey's face collapsed in shame, he hung
his head and begged forgiveness for letting the army down. Ten
months after being sworn in, he was mustered out with Honorable Mention, a category upgraded after one year to Honorable
Discharge. Mission accomplished!

From there it was to San Francisco for the Summer of
Love. But only briefly. The hedonistic hippie lifestyle wasn't for
him. Back in Oregon, he enrolled at Portland State on the GI
Bill. This proved another brief stopping point. The life of the
mind that he had so idealized wasn't to be found in academia.

Professors struck him as stale pedants, more intent on criticizing the great thinkers than celebrating them. Harvey dropped out.

Which brings us to the arc of the mythic hero, a role Harvey fulfills within the Burning Man community. According to Joseph Campbell, the eminent expert in such things, the arc begins when a seemingly ordinary person, e.g., Larry Harvey, grows intolerably distraught in the ordinary world. The polarities in his life pull him in contrary directions until, unable to tolerate relentless cognitive dissonance, he sets out on a course of his own choosing, from which there is no turning back. "Fabulous forces are there encountered," writes Campbell, "and a decisive victory is won: the hero comes back from this mysterious adventure with the power to bestow boons on his fellow man."[4]

Post–Portland State emerges as the part of Harvey's journey where he wanders into the wilderness. His version was a tiny Oregon town, Coquille (population: three thousand), where he lived with his girlfriend. While she taught elementary school, the breadwinner, he gorged on histories of classical Rome by the likes of Tacitus and Livy. "Close to paradise," he describes Coquille. Lush wetlands with beaver ponds, bears feeding off blackberries, fingerling Coho salmon fattening up for their run to the sea. He'd never been anywhere like it.

After several years, though, the appeal of small-town living dimmed. The closest library was a long bus ride away. Harvey craved distraction. In 1978, the couple moved to San Francisco, shortly after which they separated. "I was suddenly faced in my thirties with the most embarrassing task of trying to support myself—which I had never done before."

Meanwhile . . .

A legendary Bay Area bohemian named Gary Warne was

teaching a course at San Francisco State's free Communiversity. Practical Jokes 101 was a surrealist curriculum of parodies and stunts to confuse ordinary civilians. Like several dozen participants lined up shoulder-to-shoulder panhandling as if they didn't know each other. In the 1977 catalogue, Warne posted a new offering. The name came from the title of stories by Robert Louis Stevenson (*Treasure Island, Strange Case of Dr. Jekyll and Mr. Hyde*), a detective series about one Prince Florizel of Bohemia who infiltrates a secret society of people intent on taking their own lives. Warne called this class the Suicide Club.

San Francisco State would disown the Communiversity shortly thereafter, but by then word-of-mouth had launched the Suicide Club into San Francisco's underground as a secret society that anyone could join. "Members must agree to enter into the real world of chaos, cacophony and dark saturnalia," read the prospectus, "and they must further agree to live each day as if it were their last." The club endured for five years, its stated mission to explore the untraveled experiences of life. Those experiences entailed such goofs as a tour of the Oakland sewers, an overnight picnic in an abandoned hotel, infiltration of a Werner Erhard est seminar and a Moonie commune. One epic bit turned an elevator in the posh Sir Francis Drake Hotel into a make-believe shower filled with sudsy, naked Suicide Clubbers who feigned embarrassment when unsuspecting guests entered.

John Law, destined to be one of Burning Man's originals but who would later leave in an acrimonious split with Harvey, joined the Suicide Club. Eighteen years old and new to San Francisco, he was among a band of nudes who took over a cable car, an image rendered famous in a "Welcome to San Francisco" postcard. "It scared the fuck out of me," Law recalled. "My stomach was tied up in knots. And then when we did it, I was enormously relieved. Who gives a shit if I'm naked."[5]

In his confessional, "Cosmic Cosmology," Warne described his own obsession with crossing self-imposed boundaries. The transformative moment came when he persuaded a few others to join him in a climb up the Golden Gate Bridge. As he rose, he was "immersed in images of falling thru space into the ocean." By facing down those catastrophic premonitions, he came to understand the nature of fear itself as "a freeze on the future, the filter or floodgate that stops our imaginings; something within us that stops us from becoming more powerful and loving."[6]

The club distilled its ethos into a set of Chaotic Principles: Divest Yourself of Expectations; Allow People the Validity of Their Own Emotions. Number eleven in the list stakes out the imperative that "nobody spectates, everybody participates." We Subconsciously Believe We Have Experienced Things When We Have Only Watched Them. We Have Not![7] Years later, variations on this theme would be inscribed as Burning Man's Ten Principles.

The club dissolved in 1982: "After six years, everybody was sleeping with everybody else, and people were bored." A year later, at age thirty-five, Gary Warne was dead from a heart attack. Mourners poured his ashes into the waters below the Golden Gate Bridge.

Remnants of the Suicide Club morphed several years later into another distinctly San Francisco assemblage, the Cacophony Society, the mother of all "guerrilla/gorilla" underground theater. Distraction, disorder, contradiction, culture jamming—"noise"—were its watchwords. Hence the name Cacophony. Its intellectual roots traced back to Dadaism, the über-antibourgeois art movement remembered for Marcel Duchamp's porcelain urinal (rejected by a tony exhibition at the time of its creation, later deemed the twentieth century's "most influential work of modern art").[8] For Dadaists, as would be later true at Burning Man, the process of creation—not the produced object—defined art.

Cacophonists also drew from a critique of capitalism by a cadre of artists and intellectuals who called themselves Situationists. Active from the late 1950s through the 1970s, they argued that the ever-broadening spread of commercialism was depleting culture of its vitality. Their antidote? Eliminate passive spectacle, wherein the audience pays to consume "the triumph of commodities." Replace it with Situations, participatory moments deliberately constructed to reawaken desire, engender adventure, and provoke freedom. Happenings. This litany, too, would show up at Burning Man.

If the Suicide Club precursor was proto-punk, the Cacophony Society played to irony and wit. Their acts of sabotage took on surreal trappings. Gluing toasters to light posts. Crashing dog shows in dog costumes. Replacing the stuffing of dolls on store shelves with cement and attaching a warning tag ("Unfortunate child, beneath my plush surface lies a hardness as impervious and unforgiving as this world's own indifference to your moral struggle"). The karma of the movement was inclusive; they extended an open invitation to join their street theater of the absurd. "You may already be a member" was the society's tagline—blanket encouragement to become a Cacophonist and put yourself "on the knowing side of the joke."

Stuart Mangrum, Burning Man's director of education these days and Larry Harvey's longtime intellectual wingman, was drawn into the Cacophonist vortex. He had grown up in Los Angeles, "too smart for my own good," the son of a librarian mother and salesman father. At sixteen, he entered the University of Southern California but quickly quit. "I wanted to do something that was off the map. All my friends were turning into lawyers, I wanted some adventure."

He found adventure as an Air Force intelligence officer, flying over China, Korea, and Russia in spy planes as the on-board airborne cryptologic linguist (job description: process, exploit, analyze, and disseminate signal intelligence). That is all he'll say, even decades later, about this top-secret work. The only part of his résumé that wasn't classified got him hired as a corporate trainer in San Francisco. Which segued, with the birth of the Internet, into freelance e-commerce marketing gigs. And the Cacophony Society.

Mangrum, the Cacophony house thinker, produced the society's own set of principles, framed as a 12-Step Program for those aspiring to recover from their addiction to a life of banal humdrum. The process began by accepting the city as a playground; the final stage was to "leave the world a weirder place."

The Zone Trip was the cornerstone of Cacophony, excursions occurring beyond habitual coordinates of time and place. Far, far outside the comfort of routine. The idea came from a Russian sci-fi movie, *Stalker,* that portrayed a realm where secret wishes were fulfilled. Cultural critic Hakim Bey gave the concept intellectual credibility as a poetic fantasy, likening it to an insurrection that temporarily liberates enslaved territory (physical or mental) and then disappears before the state can crush it. More foreshadowing of Burning Man.

Cacophony Zone Trips were mysterious by design. A notice

would appear in their newsletter *Rough Draft*, designating a meeting spot but little more. For example: "An intense event scheduled for October 31. Not for the squeamish nor those who think that Gary Warne has been reduced to ashes. Limit of twenty by application." Daring folks showed up not knowing what awaited. The announcement in 1990 for Cacophony Zone Trip #4 gave more details than usual: "We shall travel to a vast, desolate white expanse stretching outward to the horizon in all directions . . . A place where you could gain nothing or lose everything and no one would ever know. A place well beyond that which you think you understand. We will be accompanied by the Burning Man, a 40-foot tall wooden icon which will travel with us to the Zone and there meet with destiny."

Meanwhile, long before Zone Trip #4 . . .

Larry Harvey was finding his footing in San Francisco. For a while he sold hot dogs at the Farm, a cavernous rock club (the largest punk venue in town, also the most violent). He drove a cab. He briefly married a Jamaican woman; they never lived in the same house but conceived a child, Tristan. "My son was the perfect miracle to me," he told a reporter in 2014. "Being adopted meant I'd never met anyone genetically connected to me. My son had this look as a baby, a sort of 'You will respect my boundaries.' People who came up to hold him would get this look and I thought, 'Oh my god, that's the look I've given people my whole life.' It was so deeply affirming and reassuring to see it. Though of course he wasn't me at all; that's the big mistake that parents make."[9]

A small inheritance from a friend of his grandmother allowed him to buy a truck to start a small landscape business, the job cited when reference was made to his occupation before

Burning Man. Paradise Regained was its name, borrowed from John Milton's sequel to *Paradise Lost* ("I who e're while the happy Garden sung, By one mans disobedience lost, now sing Recover'd Paradise to all mankind"). Like Milton's epic, Harvey conceived of life as a song of awaiting redemption. His signature landscapes weren't tidy beds of plantings but wildly inventive installations that combined flora with incongruous objects (he bordered one client's yard with bowling balls). "I wanted to create Mars!" His work led him to a troop of intellectual craftsmen, "latte carpenters" they called themselves, blue-collar guys with cultural chops. "I really wasn't an artist. I was hanging out with these famous latte carpenters, all of whom, in their spare time, were writing novels or painting pictures or playing music."

Around then, the agonizing split with the girlfriend happened. An ardent autodidact, Harvey spent large chunks of three post-breakup years in bed reading psychology texts in hope of rousing himself from depression. Conducting self-therapy. At first, there was Freud. "Each night I'd look down at his photo on the inside book flap for ten to fifteen minutes, I'd see this benevolent-looking figure with white hair and he became the father I didn't quite have. It was classic transference."[10] But it would be the work of Heinz Kohut, who pioneered the study and treatment of narcissism, that most resonated. That I knew of Kohut from my own psychological journey surprised Harvey that day in his flat, as he sipped water from a coffee mug and puffed away on nonfiltered cigarettes. It constituted an intellectual bridge between us, encouraging him to talk to me as he might a therapist. "Real love entails authentic care," Harvey mused. Then, with pensiveness appropriate in a confessional, he talked about his black night of the soul. "She had her own set

of issues but I was too stupid and selfish to see them. That made
the wound all the worse."

Finally, by the spring of 1986, he'd had enough self-pity.
One morning, as he tells me the story, he awoke without awful
despair in his belly. He wasn't yet out of the woods but, having
briefly tasted how good it felt to be without that ache, he became determined to break the downward cycle of misery that
held him captive. A beacon of light inspired him, the memory
of the night he and his girlfriend built a campfire on the beach
to celebrate the summer solstice. Young Tristan was with them,
squirting lighter fluid on the ground and igniting it with a glowing poker. The trio knelt together, a primal family, setting alight
patterns in the sand. "It was the most romantic moment I could
recall."

How better to cauterize his pain than by repeating that experience? "I was tired of passively suffering, of feeling empty, of
feeling soul-less." And with another solstice imminent, he called
up his carpenter pal Jerry James to suggest they build something
and go down to the beach to burn it. Even from Harvey, the
request was so . . . utterly random . . . that James made him repeat
it. They ended up nailing scraps of wood together with a trellis
at the top, materials at hand in Harvey's gardening workshop. It
wasn't meant to represent anything. The sole design requirement
was to stick upright in the sand. Harvey has often wondered
how that stick figure arose. He got that distant look in his eyes
when I asked. "Who knows where art comes from?"

Whether or not the break-up had anything to do with that
original Burn has, by now, been lost to lore. He would, when reporters asked over the years, downplay and then ultimately dismiss his broken heart as the origin of Burning Man (just as he

would dismiss suggestions that the first construction was meant to embody his former girlfriend or her new boyfriend). An early student of Burning Man, however, insists that Larry specifically told him he wanted to purge her from his memory. "He moved off that soon afterward," the author wrote. "I guess he figured it wasn't very politically correct, and now that idea is actively suppressed."[11] On the other hand, maybe he didn't want to keep rehashing dark times. Regardless, as Harvey chain-smoked three decades later, serene in his recollection, this was the version he recounted to me.

A typical dark, foggy, windy San Francisco night is how Jerry James remembers the evening that he and Larry and their boys held the first Burn. "Besides our little group, there were really only about four other people around, and they came over when we ignited this thing. One of them had a guitar or a tambourine, and that was sweet. I think they wanted to get warm."[12]

Befitting his naturally lyric mode, Larry recalls the event in more grandiloquent terms. "Strangers ran and joined us. Suddenly the crowd doubled or tripled. I could see everyone's face lit by the flame. We were moved as one is moved by the enthusiasm of strangers for something you've done. Something about the kindness of strangers, unconditional love. It was touching. Those acts of impulsive merger and collective union made the evening so special. I'm very much of the conviction that we would never have done it again if those circumstances had not happened and helped us be so moved by what we'd done."[13] Which sounds a lot like what he hoped would happen in that corn maze.

Harvey and James decided to stage the same party the next year. This time, they handed out flyers in advance advertising the event: "join us at the northern naked end of baker beach on

Saturday, june 20 for the second annual burn." The Cacopho-nists got word of it; a happening like this was right in their sweet spot, a modest Zone Trip. A few of them showed up. "The first one was like a family picnic," Jerry recalled. "And the second one was like a bigger family picnic."

For the third one, Harvey gave the effigy and the event a name: Burning Man.

2

Gravitational Waves

The newsletter in 1997, it's one big propaganda sheet for community. That's when the litany began—community, community, community. People grabbed onto it.

LARRY HARVEY

"I've always been supportive of Larry's vision that we are going to change the world."

The year is 2016 and events have advanced so far in the intervening decades that such a remark, that Burning Man might change the world—throwaway line though it may be—is credible enough to take seriously. Not only the remark but also the speaker is testimony to the evolution of Harvey's bonfire. Marian Goodell heads what has now morphed into the Burning Man Project, a \$45-million-a-year nonprofit. The audience is further testimony, comprised of delegates from over two dozen countries representing events and communities that owe their inspiration to that summer solstice Burn.

In a black-and-red dress that complements her curly blond hair, Goodell is enthusiastically sharing with the crowd the collective faith that they are creating something much bigger than a global party. She's the Burning Man Project CEO, chief engagement officer. Her buoyancy in reaffirming Burning Man's seriousness, its lightly worn gravitas, aptly reflects the job title.

She is addressing the Global Leadership Conference, representatives of the loose confederation known collectively as the Regionals. Throughout the world, in locations diverse as South Africa, Japan, the United Arab Emirates, Australia and New Zealand, Argentina, throughout Europe, even Shanghai, are pockets of Burners. Gathered amid the splendor of a gilded hotel ballroom in San Francisco, they are testimony to Burning Man's influence in unlikely places. The various events they represent are not subsidiaries or corporate outposts but rather independent, autonomous, self-organized variations on a theme.

The Global Leadership Conference is their annual networking exchange. It's been meeting since 2007 to share experiences, trade war stories, and pool knowledge. But this one is different, in mood and context. The free-for-all life force traditionally at the heart of these gatherings hasn't diminished but, ever so subtly, a movement is afoot to coax all that raw stem-cell energy into something more . . . manageable. And permanent. There's a subtle subtext to this session, allusions to the latent potential inherent in the organization's new iteration as a nonprofit, the Burning Man Project. There's now more basis for taking the rallying cry seriously. Change the World!

To that end, Marian was announcing big first steps the Project was taking, adding two high-level positions. A new tier of bureaucracy. Despite the culture's long-standing predisposition to anarchy, its disdain for central authority, its pride in do-it-yourself ingenuity rather than top-down direction, headquarters was adding two professional managers from the outside.

A director of development was now on board, "to execute fund-raising which celebrates Burning Man's culture of gifting." In other words, raise money without selling out! And a director of art and civic engagement to oversee expansion of the community-outreach sides of the operation. Which is to say, "collaborate with artists and participants in the creation of life-affirming interactive experiences." The Burning Man Project was re-engineering itself, adding administrative heft and firepower.

That's how far things had come since Harvey arose from that emotional stupor in 1986. The organic phenomenon had become institutionalized. And nobody in the audience yet knows about Fly Ranch.

Four years earlier, in 2013, Burning Man had changed legal form to position itself for bigger things. Previously, it had been a limited liability corporation, Black Rock City LLC, owned by six people known collectively as the Founders. An early article about these main movers described them less than charitably as "a group of slackers who grabbed hold of the one thing that brought them notice." An unemployed landscaper (Larry

Harvey), an art model (Crimson Rose), a struggling photographer (Will Roger Peterson), an aerobics instructor (Harley K. Dubois), a sign maker (Michael Mikel), and a dot-com PR gal (Marian Goodell) "who made good."[14]

Make good they had. Which, as an unintended consequence, forced them to think about how to ensure continuity after they were gone.

The LLC took its name from Black Rock City, the annual temporary settlement in the Nevada desert of seventy-plus thousand people who show up with tickets for Burning Man. A lot more, easily twice as many, would eagerly come, but the population is sealed by edict of the Bureau of Land Management (BLM), the federal agency that manages the land. Frenzy for those tickets is intense. Within hours of the Internet site going live, general admission sells out at $390 apiece (with another $80 for a vehicle pass). On the eve of the event, the going Stub-Hub price tops $1,400.

Despite such demand, Burning Man's popularity is still on an upward trajectory. In the first years, the bulk of attendees came from the West Coast, especially California (San Francisco so empties out that reservations at primo restaurants are easier to get). Of late, though, a growing portion comes from elsewhere in the U.S. and other countries. There remain lots of areas for growth, particularly east of the Rockies, where in cities like New York or Boston the mention of Burning Man typically draws a blank stare. Those who *do* recognize the name are likely to sneer disdainfully, knowing it only from a sliver of unsavory media bites. But that is changing.

As Burning Man has grown more popular in recent years, marketers spotted its commercial appeal. Among millennials, the brand exerts a heavy appeal, conveying an image of hipster cool—which makes it hugely valuable. Various entities, including some high-tech household names, took a pass at buying the LLC. Trial balloons were floated at massive multiples of book value. Had the Founders wanted to get rich, had they sought to monetize their creation, there was ample opportunity. But, backbiting gossip that they were becoming wealthy to the contrary, the overtures fell on deaf ears. Cashing out was never their exit plan.

Not to suggest that "these slackers" were pure and saintly. No professional do-gooders, these folks. Indeed, one apostate claimed Harvey's persistence in keeping Burning Man alive in the early days was so he wouldn't have to find a real job. The motives of the LLC were a mash-up of cerebral humanism, mischievous mayhem, and the pleasure of staging a most unlikely event with your pals. Building an anti-consumerist playground was an end in itself, a playground that kept expanding in size and complexity. It kept them endlessly engaged. Their agenda wasn't net worth.

The sprouting of the Regionals was what got the LLC to fully appreciate the potential tiger it had by the tail. All those folks in the San Francisco hotel ballroom. With four of the six Founders in their sixties and seventies, still spry but ever-longer in the tooth, matters of succession needed to be addressed. If their intention to change the world were to become more than idle fancy, more than a doper's pipe dream, institutional foundations were required.

The event would always remain the keystone, the mother ship, but the leadership's gaze was surveying a more distant horizon. The Founders started thinking in a more extended

time frame, contemplating possibilities that would take a long
time to unfold. At headquarters there was informal talk of a
"one-hundred-year vision."

The LLC couldn't provide such institutional continuity, not
with six owners each exercising equal authority (to say nothing
of what would happen when equity passed on to heirs). If Burning Man were to be around a century hence, the corporate apparatus needed to become more agile. Management equally had
to become more adept. Bigger staff meant bigger budgets. Gate
receipts were paying for the entire operation—with remarkably
little left over from year to year to bank as capital. Which meant
identifying new revenue streams. Becoming an agent of historical change doesn't come cheap. The kind of social transformation being imagined made it imperative that the Project find
additional financing.

Michael Mikel was the first Founder to advocate a change
from LLC to nonprofit. He was the Cacophonist who discovered Larry Harvey's third beach Burn and brought in the others.
On that first trip to the desert, he famously dug a line in the
sand with the toe of his boot and declared that all who crossed
over were about to enter a new world. It was he who went off
into the immense playa at night to locate lost folks in those early
days when there was no trail into Black Rock City from the
highway and attendees had to navigate by compass and odom-
eter; a wrong calibration meant circling around endlessly in the
treacherous badlands. Out of his search-and-rescue missions
came Burning Man's Black Rock Rangers, the community volunteers charged with order keeping and safety. And his iconic
Danger Ranger playa persona. Over the years, he reigned as

Burning Man's courtly tribal elder, "silvery, lean, with eyes of pure charm and mischief."[15]

The other Founders understood in the abstract the need to reorganize to ensure continuity. There was resistance nonetheless. Surrendering control was tough. Agreeing to this new institutional structure took three years: endless meetings, conversations that never seemed to reach conclusions, flare-ups, and personality clashes. A therapist was even brought in to smooth the waters and soothe injured feelings.

The contours of the new entity took shape. It would be the umbrella sheltering the different aspects of the whole shindig, some of which had been operating independently. Under the LLC there had been a board that was, at best, advisory. The nonprofit, on the other hand, would be presided over by a bona fide board with legal authority and responsibility for decisions. The Founders could no longer rule as benevolent grandees.

Removing them financially posed a monumental dilemma. At what price? The appraised value of Black Rock City LLC was $7.3 million, representing $1.2 million to each shareholder. The Burning Man community, notoriously opinionated and contentious, lay in wait for anything resembling a sweetheart deal. "Whatever was paid had to be culturally appropriate," cautioned Kay Morrison, an LLC board member from Seattle who would also sit on the nonprofit board. "If they got millions, people would say 'they were in it for the money.'" In the end, each Founder accepted $46,000 for their shares, a pittance of their worth.

At the eleventh hour, with the finish line in sight, a wholly unanticipated wrinkle threatened to scuttle the whole deal.

Who would control the brand trademark? Some years earlier, the Founders had set up a second entity, De-commodification LLC, to own the Burning Man name. This arrangement had never been publicized. Much of the board, in the dark about this, was understandably stunned. If the Project wasn't going to own the name Burning Man, what exactly *did* it control?

The Founders, on the other hand, were reluctant to surrender the trademark. They proposed continuing the long-standing contractual arrangement whereby they would license it each year "for not more than $75,000." They worried about handing over their brand to an untested entity that might crash and burn. Should that happen, if the Project failed and dissolved for any reason, California (where the Project was registered) would be empowered to transfer all assets, including the name, to another nonprofit of its choosing. The Founders' caution was justified. "We didn't know if the nonprofit could hold it together," conceded Kay Morrison. The Project was slated to start life with a mere fifty-thousand-dollar bankroll and accountability for a multi-million-dollar operation. "It was like a kitten taking on a monster."

A concession was offered, a kind of insurance policy whereby De-commodification would own the trademark for five additional years, then hand it over to the Project "to be held in perpetuity by the nonprofit for the benefit of Burning Man Culture," details to be determined at the time of transfer. For some board members, though, even that wasn't good enough.

The clock was ticking. There were consequential tax advantages to finalizing the deal by year's end 2012, which lent urgency to the negotiations. To further exacerbate matters, the board learned on the eve of the vote that organizational bylaws required unanimous consent—not, as had been thought, a simple majority. On the evening of December 27, with the outcome very much in doubt, the board gathered in San Francisco to cast ballots. Kay Morrison wasn't among them. She was home in Seattle, participating in the proceedings via computer.

Morrison's youthful emergence as a respected figure—she was in her early thirties—testified to her stature in the community. A veteran human rights activist at a young age, she first heard about Burning Man in 1999 from a fellow demonstrator at a big protest against the World Trade Organization in Seattle (where she spent four days in jail). She had planned to be in Prague the next summer to demonstrate against the International Monetary Fund, but found herself banned from the Czech Republic as a result of the WTO arrest. At the last minute, with uncommitted time on her hand, she went to Burning Man on a whim. Uncertain what awaited and pretty much unprepared. To add to her woes, the weather that year was awful, rainy and cold.

In most every respect, her experience was terrible. She likely would never have returned were it not for an epiphany at the porta-potty. Among the few things one gets with a Burning Man ticket is access to public toilets. Hundreds of them are strategically positioned throughout Black Rock City but there are never enough. Kay had to patiently wait her turn. "Standing in line, everybody was so warm. So chatty and familial. Nothing like I would have expected. The flash of that moment got me thinking about what could happen if you broke down barriers. Could you have these kinds of interactions all the time? I was hooked."

Over the ensuing years, she became a leader in her local

Burner scene, playing an important role organizing the regional Critical Northwest. Professionally, she worked as an administrator in nonprofits. For recreation, she was a blacksmith artist with the Iron Monkeys Metal Collective, whose pieces got displayed at Burning Man. With her commitment to the community so visible, she was named to the LLC board.

As the voting started, with puppy Blaze beside her on the couch, Kay sat on pins and needles. All day she had worked the phones, talking to the holdouts, coaxing and cajoling, seeking a way forward. "I regard myself as glue, sticking people together." In effect, she was asking each side to trust the other. The calling of the roll began: "Do you vote yes or do you vote no?" Kay poured herself a tumbler of whiskey, poised for celebration or remorse. Either way, she'd need a drink. "When it came in 'yes,' I crumbled."

The Burning Man Project legally incorporated a few days later, a 501(c)(3) nonprofit capitalized with the donated LLC shares of the Founders. Its outward-facing intention was formalized in the legal mission statement: to provide infrastructure, art programs, and educational resources to apply the principles of Burning Man in many communities and fields of endeavor. In a word, to spread outward. "After twenty-four years of tending our garden in the desert," Harvey proclaimed at the announcement, "we now have the means to cultivate its culture worldwide." The pump was primed.

The genius of the Founders had always lain in how they functioned as a unit. Their whole was, indeed, so much greater than the parts. "The six of us were married to each other all those years" was how one put it. "We learned that we made better decisions as a group than individually and came to lean on each other." Taken as a whole, they were a remarkable assemblage of complementary talents. Harvey was the philosophical seer and provocateur. Michael Mikel was sagacious and supercompetent, a straight talker who inspired confidence. Will Roger took charge of building the city, laying out roads and erecting structures and creating Burning Man's first salaried department, Public Works. Crimson Rose was an aesthetic spirit who turned what might have been mere activities into elegant ritual through form and dance. None, however, because of age or temperament, was a candidate to run the Project.

Marian Goodell and Harley Dubois, a generation younger than the others, were the logical choices. Harley had soft people skills par excellence; she gets credit for masterfully crafting and nurturing the volunteer system that enables Burning Man to work. Marian had the temperament and savvy to make good business decisions; she networks deftly and appreciates the value of positive PR.

With their different styles, compounded by the fact that each was a former Larry Harvey girlfriend, the two women had been each other's nemesis in power struggles over the years. Choosing a leader was not going to be easy.

Possessing an off-the-charts emotional intelligence, Harley had the superpower to see things that need doing and organize their undertaking with a light touch that engenders maximum goodwill. That morphed into her role as Black Rock City Manager, the woman who kept the trains running on time. Descended from a line of ministers in Canada and South Dakota, she contributed to the Founder mix a steady moral compass. "My true skill lies in understanding and loving people."

As Burning Man grew, diverse factions had to be reconciled.

Harley's talent for resolving conflict became ever more conspicuous. Her personality made her a natural consensus-builder,
the house statesman able to negotiate treacherous waters by patiently getting buy-in from the right people.

Marian brought different talents to the team. She, more
than any of the others, was comfortable taking care of business.
Her affable personality made her good at networking; she drew
people of power and influence into the fold. Family background
and East Coast experiences had taught her how the game is
played in the big leagues. She could sometimes be perceived as
harsh, as saying no prematurely. But her instincts were shrewd,
her commitment to Burning Man unequivocal, and her heart
warm.

Things might easily have fallen apart as talks ensued, the
nonprofit model collapsing before it even began, had there been
an insoluble impasse over which of the two women would head
the Project. Each had her camp. "It depends on you," the con-
sulting therapist advised Harley at a tense moment in the process. "If you can tell me what you want to do, the others will
follow." Harley felt the pull of her family, her husband and ad-
olescent daughter, of her desire to be present for them. Accepting Marian as First Among Equals, her de facto boss, might be
humbling, but in answer to the psychologist's question, that's
what Harley finally decided to do.

On a mild morning in the spring of 2016, Harley and I sat
on the terrace of a North Beach café. Her six-month-old puppy,
Hazel, was leashed under the table, restless for its daily walk.
This was our first meeting; she didn't know what I looked like
so had come early to post herself conspicuously by the entrance.
Her vibrant, robust personality immediately kicked in, as did a

whirling mind barely able to form words fast enough to keep up with her thoughts.

The nonprofit at this point was several years old, its organizational chart growing more differentiated, the directors of development and of art and civic engagement now on board. Harley expressed no regrets about becoming number two, by title chief transitional officer. What did her role entail? "Mostly sitting in meetings, mentoring senior staff, instilling confidence with reassurances of 'yeah, you're doing it right!'" But at fifty-three she was restive, telling me how appealing it would be to become a full-time mom. Or to resume her interrupted career as a painter (as a student she had gone to art school in France). "I feel fortunate to have been here but I don't need this. Burning Man doesn't say who I am."

Marian was treating her with "respect" and "flexibility." All systems were go. Should she choose to leave, it would be on a good note. I didn't need clairvoyance to see the ambivalence: part of her wanted to move on, another portion wanted to remain as an integral piece of whatever this new thing was to become. Which, although I didn't know it as we spoke, would include Fly Ranch.

The mere possibility of a gathering like the leadership conference of 2016, or a sprawling network of kindred spirits who regarded Burning Man as their true north, was a zillion-to-one shot back in the beginning. Equally improbable was the presence of Marian Goodell on center stage addressing them, making comments such as "We're scaling to meet the growing demand for tools and resources to reproduce the Burning Man experience."

That Burning Man experience, while not entirely

unrecognizable from those first happenings on Baker Beach, had metamorphosed into something wholly different in countless ways. For starters, it had left San Francisco after the Golden Gate Park police prohibited the burning of what, by 1990, was a full-blown forty-foot effigy. Too dangerous, authorities said. The big crowd too unruly. Harvey was distraught, watching his reverent tribute to the noble human spirit threaten to devolve into "a mere roadside attraction."

By then, there were lots of people who wanted this . . . thing . . . to continue. That quintessential San Francisco boho crew, the Cacophony Society, was now fully integrated into the scene. The cops couldn't stomp out the sparks. After some Northern California sites were ruled out, Nevada was suggested. Several Cacophonists knew about an isolated stretch of federal land northeast of Reno. Where they went to fire guns, shoot off explosives, and drive cars super-fast. With no other real option, Harvey agreed. Sight unseen. In a migration recalling the Mormons' journey to the Great Salt Lake, an entourage embarked for the desolate Black Rock Desert. Zone Trip #4.

"I was struck by the enormity of the moment," recalled one of those original pilgrims. "We were packing up, getting ready to leave the city to do what? It didn't matter. It was 'the big A'— an adventure into the unknown. It was more than leaving the city culture and shedding our urban upbringing. Eighty people had suddenly become eighty friends who cared for each other, had made a commitment to follow this strange wooden statue to the desert and live with it till its last moments when wood was to become ash and smoke."[16]

The Black Rock Desert is an austere, intimidating terrain of mountainous rock straddling narrow gorges. The early pioneers passed through there in ox-drawn covered wagons, the fabled prairie schooners. The principal track westward went through this territory, and a bit farther the road forked with one branch leading to California and the other to the Oregon Trail. Then, as now, this was tough, unforgiving country. If a wheel broke and there was no spare, the wagon was abandoned. To this day, access routes are unmaintained, dirt roads require vehicles with a high clearance, water is scarce, and cell phones don't work.

The topography is primordial. Ancient lava beds poke through the crust of the land, producing hot springs often too fiery for safe soaking. The vista extends in a succession of ranges and basins—horsts and grabens to geologists—natural fault lines where the earth sinks from its own weight between granite uplifts.

Eons ago, when weather there was wetter and cooler, a massive inland sea of some four hundred square miles existed in one of these enclosed basins. Some nine thousand years ago that water disappeared as the climate became semi-arid. Fossils of ichthyosaurs, sixty-foot-long "fish-lizards" that slithered along the bottom of the seabed, continue to be found. When the first white trappers arrived, descendants of the indigenous inhabitants still resided in the region, a people calling themselves Agai Ticcatta, "trout eaters," in remembrance of the oversized fish that once abounded here. A surveyor sighting the route for the transcontinental railroad gave the dry lakebed a name, Lahontan, in honor of the French explorer who commanded Fort Detroit in the 1680s.

That ancient lakebed today is a vast expanse of cracked earth stretching as far as the eye can see. The Air Force used this hardpan to practice aerial gunnery during World War II.

It was here that a car, propelled by two Rolls-Royce fighter-jet engines, clocked a world record of 764 miles per hour in 1997, the first time a land vehicle broke the speed of sound. Gypsum proliferates in deposits from the antediluvian lake. The local settlement of Empire, where all the houses were owned by the United States Gypsum Corporation, used to manufacture it into sheetrock. When the plant closed in 2011, a victim of cheap Chinese drywall, the workforce went from 95 to 2 and the town lost its zip code. Gerlach (population 206 at the last census) is the regional hub; a sign on the outskirts marks the spot where "The Pavement Ends and the West Begins."

Most months of the year these 300,000 acres are bone-dry. Temperatures hover around 100 degrees by day and chilly at night. At first glance, the landscape looks lifeless. No bugs, no birds, nothing green. Dust storms sweep the plateau clean. During winter, when most of the annual seven inches of rain falls, the flats flood ("flood" being a relative word for what elsewhere would be puddles of standing water). That's when microscopic phytoplankton and bacteria appear. As do tiny water crustaceans from buried eggs that hatch in the runoff: water fleas and three-eyed tadpole shrimp whose lineage dates back seventy million years. Come spring, the mud dries. It leaves in its wake, according to Department of the Interior geologist William C. Sinclair in his 1963 study *Ground-Water Appraisal of the Black Rock Desert Area*, "a hard clay surface glazed with fissures and white with the efflorescence of minerals precipitated from the saline water."

This is the playa.

The playa is an apt representation, a graphic metaphor if not a physical facsimile, of the planet before the Garden of

Eden. Imagine it as the horizontal axis from which life on earth sprouted. "The potential for agricultural development," concluded Sinclair's report "is limited by the arid climate, the steepness of the alluvial slopes flanking the mountainsides, and the salinity of the soils and water." A wasteland. Since 1990, this has been the locale for Burning Man's civic polis, the ongoing social experiment in creation and deconstruction called Black Rock City.

Back on the San Francisco stage, Marian was explaining to the global leadership how the Project was preparing to accelerate the spread of Burning Man culture beyond the austere terrain of Black Rock City. Such an objective, reasonable though it is, would have sounded positively mad in the mid-'90s when Burning Man's survival itself was at risk. The imperative back then was simply to stay alive. Marian was among those holding on to the ledge by their fingernails.

Her tale, starting with her first Burn as one of several thousand attendees in 1995, had a considerably different arc than Larry Harvey's. Born in Baltimore into a well-off Irish Catholic family, she was raised in the northwest corner of Ohio in the small town of Bryan (population eight thousand), home of Dum Dums lollipops (with such flavors as buttered popcorn, s'mores, and funnel cake) and Etch A Sketch. Her father, an MIT mechanical engineer with a Harvard MBA, managed a rust-belt brass and copper smelting company. Her mother was not only a Wellesley College grad but also a former Mardi Gras queen (Krewe of Mystery).

As the oldest of three girls, Marian did the babysitting and was accustomed to getting her own way. She remembers herself as a shy kid, playing clarinet in the marching band and earning so many Girl Scout merit badges that, if a boy, she would have been an Eagle Scout. Aspiration came naturally. One sister grew up to become an environmental engineer/venture capital type, the other a physician running a clinic for poor children.

The tradition in her family was for women to go to prestigious Wellesley but Marian didn't even apply. "I didn't think I'd get in." Instead, she attended Goucher, another old-line women's college outside Baltimore, where she got a BA in Creative Writing. It was there she first got a taste for leadership: elected treasurer of the student body, in charge of allocating funding to campus organizations ("my dad taught me how to manage money").

Post-grad she went to Boston. Her parents had met there and, since she planned to marry somebody like her father, the move made sense. She worked in advertising, then public relations. "I liked Boston but didn't love it." A college friend invited her to come visit in San Francisco, where Marian was infected with the bug to live in California. She returned to Boston for a few more years, working now as a legal secretary and saving money for the relocation west she knew was coming. On Valentine's Day 1988, she set out across country in a VW convertible.

Computers were relatively new at the time, and because of her familiarity with them in Boston, she found employment in the Bay Area with a tech company. Her job was teaching customers to use the equipment, which morphed into sales. She was living in San Mateo, sharing a place with two male roommates ("first time I ever lived with men"), awaking each morning to make scores of prospecting calls or spend days on the road seeing clients. But life wasn't working out how she wanted.

She had come to California seeking personal liberation. "I was certain it was going to give me room to breathe and be myself." That was the plan. Instead, she found herself thirty years old, residing in the suburbs, confined to a narrow social circle, and without serious romance. "Living hell" is her blunt description. Even worse, given the purposeful examples of her parents and siblings, she was professionally adrift. Finally heeding her father's insistent advice to find a career, she moved to San Francisco to study photography. Now she had a goal. "I don't commit easily, but when I get into something, I get all the way in."

At art school she heard about Burning Man and bought two tickets for herself and her boyfriend. Joe was also a photography student, a blond surfer-dude from LA who smoked a lot of pot. "The opposite of the guys I had been going out with when I was in sales." But they ended up not attending. At the moment of truth, they got cold feet, deciding they lacked sufficient camping equipment. Instead, they took a vacation water-skiing.

The next year, by then seasoned from a five-week trek though Baja by jeep, they were ready to confront the playa. Marian concocted a hat made from Styrofoam decorated with old-fashioned pinwheels that spun in the breeze; when someone complimented it, she gifted them a pinwheel. What surprised her most, what really blew her mind, was how . . . natural . . . she felt in the surreal setting. The constrained, cautious woman shunned her self-consciousness and jumped into the mosh pit. An unfamiliar feeling was being drawn out of her, something she suspected had long lain hidden below the surface. "I was provoked to be playful."

Burning Man hooked her. She and Joe did up their campsite in a suburban motif the following year. Green grass, pink flamingoes, a real mailbox. They poured cocktails to passersby from vintage martini shakers. And then, in one of those unexpected moments that really *does* redirect life, late one afternoon as she headed with friends toward the Man effigy, she spotted a guy in a Stetson hat. "I told my crowd to walk faster, to hurry."

Marian recognized Larry Harvey from articles about Burning Man. Indeed, she had become so fascinated by him that she had even rehearsed opening remarks on the off chance they might meet. And here was her opportunity! Catching up, she had her question ready to go: "Why the Black Rock Desert?" His answer was unmemorable, more of a tired mumble than the overture to the rich dialogue she had anticipated. He would later insist that he recalled their first encounter, but she's sure he's lying. Contact, however, had been made.

That fall, she and Joe got an invite to the Burning Man organizers' post-party. They were winding down their love affair, in that final phase where they went out with other people but protected Saturday as their date night or maybe Sunday brunch. Midway through the evening, Joe departed for another party with Marian's blessing, where he met the woman he ultimately married (to this day, Marian and Joe remain friends). Marian stayed. And ended up having that intense conversation with Larry Harvey.

It was prompted by a comment her mother had made in response to Marian's description of Burning Man. "Children teaching children" was her mother's observation, likening it to the Montessori schools where she taught. Marian eagerly sought Harvey's reaction to that analogy. "I tripped his intellectual trigger." Love at first sight? "Love at pretty-early-on sight." He was quite drunk by evening's end and asked if she'd drive him home. Their first date took place a few weeks later when he brought her to a Jerry Brown political gathering in Oakland. Afterward, perhaps surprised by her smarts, he remarked that she had "held her own." She took it as a compliment. They would be a romantic item, a couple, for the next five years.

The existential challenge to Burning Man, the moment of truth after which there was no turning back, came in 1996. The year of chaos and catastrophe. "Burning Man went over the edge," wrote Stuart Mangrum, the Cacophonist who served as director of communications at the time. Attendance, which had doubled every year since 1991, hit eight thousand. Vehicles could freely drive around, and the site wasn't surrounded by a crowd-control fence. "Too many people for the infrastructure, too many gawkers, too many cars, too many arrests, too many heartbreaks. After years of beating the odds, we got our asses whipped, and we did it to ourselves," wrote Mangrum.[17] Things were so bad that, at event's end, the BLM told the organizers they wouldn't receive a permit the following year.

Calamity cast a pall over the event even before it began. Michael Furey, "a well-known hellion in the community, a funny-dangerous firebrand who hung out with a punk-polka act," capped off a long drinking session in Gerlach. He refused offers of a ride back to camp, choosing instead to hop on his motorcycle. Which he drove onto the playa without headlights. According to legend, he headed straight for an approaching van, playing a drunken game of chicken. He lost, and was killed.[18]

Then 1996 was the year of the first art theme, Inferno, a Dante-inspired motif that proved hauntingly prescient. At the entrance to the centerpiece, the headquarters of a corporate predator with designs on Burning Man named HELCO, stood a neon Gate of Hell.

Theme camps were just starting to take hold; Harley Dubois

was in charge. "Orchestrating the budding network of this community was her baby," recalled Tony Perez-Banuet, better known as Coyote, in his chronicles. She did this from a rudimentary information board, Black Rock City's main lifeline. On one occasion, an outlandish crew wanted to torch the board for sheer amusement. Coyote watched as Harley stood her ground, dressed in a prom gown and full-length white satin gloves. "I had never seen such moxie."[19]

Harley's showdown was a mere foretaste of the bedlam that reigned the final night in 1996, the night of the Man burn. In years thereafter, the burning of the Man would be carefully cordoned off and patrolled; this night was a free-for-all unfolding to a soundtrack of automatic-rifle shots. "Explosives and fireworks were going off everywhere. One had to keep a sharp heads-up, live mortar bombs were being lobbed around like softballs." The collapse of the edifice was a "pyro popping off with huge explosions sending burning shrapnel everywhere."[20]

Tragedy was to follow that night as spontaneous, uncontrolled fires burned throughout Black Rock City, "smoke billowing with the toxic tinge of burning plastic." Roman candles were randomly launched into campsites. At dawn, a helicopter landed in mid-playa to make an emergency medical evacuation. Two people, asleep in their tent, had been run over by a stoned-out driver heading for the "rave ghetto"; the car then crashed and spilled scalding radiator fluid over another woman.

The next morning the authorities convened a meeting with the leaders to say they couldn't come back in 1997. "Enough!" said some of the organizers themselves, to whom Burning Man had become apocalyptic in more than metaphor. Among them was John Law, the quintessential Cacophonist. ("Larry was just

burning a scarecrow on the beach before he met John Law."[21])
Law was a formative figure in the early years. Relocating to Nevada had been his idea; he knew about Black Rock from racing monster trucks on the desert. Along with Harvey and Michael Mikel, he owned the rights to the name Burning Man. Even before 1996, he worried that the event had grown too big. In his mind, it was never intended to attract more than, say, a hundred. It was impossible to restore the site to pristine condition. A whole new kind of attendee was coming, the *outré* vibe of the originators was giving way to a tamer imitation.

Now, with the death of his friend Michael Furey and the chaos of the event, Law had had enough. "We were at a crossroads," he told me. "It was either going to be a giant thing or it was over, and we had no capacity to facilitate growth." At the meeting with the authorities, he took it upon himself to declare that Burning Man would not be back—that it was done. Later that day, he had a colossal fight with Harvey and then walked off. Never to return.[22]

Not all the organizers, however, felt the same way. As they huddled among themselves in the aftermath of the meeting, their circle enveloped in "the pall of shock and grief," Harvey argued for going on. But not as before. "Either we don't do this again or we drastically change our ways." The city had grown too big. Anarchy had lost its enchantment. Black Rock City had become dangerous and needed to be able to protect its citizens. In a word, Burning Man needed rules of civic engagement.

The no-holds-barred days had to give way to something more . . . civilized. Innuendo and politics cleaved the community, ugly things got said. Stuart Mangrum followed John Law's example and withdrew from the organizers (although he, unlike

Law, would continue to attend the event). He unleashed his pen against Harvey, describing him as somebody who "really wants to be a genius, and maybe he deserves to be, but passion and rhetoric are no substitute for substance." The two would eventually reconcile; years later Mangrum became Harvey's cohort in picking the annual theme, but in the passion of the moment he saw only feet of clay. "In another era, [Harvey] might have sold patent medicine and done pretty well at it, only occasionally ending up in the tar and feathers."[23]

"Inclusion" was Harvey's counterargument to those who wanted to fold up the tent. Go big! That was his mantra. The more the merrier! "That was, for me, a redemption," he said about the waves of people drawn to Burning Man, "to make a society that everyone could come into. It was an antidote to the alienation I grew up with."[24]

Maybe Mangrum was right, maybe Harvey was blowing smoke, but he continued to insist something was at stake that was grander than revelry. There was *power* embodied within Burning Man, pure quanta able to energize thought into action and then back again. Articulating this wasn't easy but the folks on his side of the argument seemed to understand what he meant when he talked about the event as an enabler of "human-ness." About how it brought forth creative essence. How it facilitated so much interdependence among strangers that one might dare say it inculcated a collective soul. Don't get me wrong, Harvey would concede, the ultra-freedom ethos of the Cacophonists remained compelling, but Burning Man was now about much more than raising hell. If rudimentary rules were the price to pay for realizing its latent potential as a matrix within which people could liberate themselves, so be it.

"Are you with me?" he asked Michael Mikel. The choice wasn't easy. John Law was Mikel's best friend and his partner in a neon-sign company. He shared many of Law's qualms about Harvey, with whom his relationship would become strained in later years. Still, he chose to stay. "I believed that what Burning Man was doing was for the greater good and decided to persevere."

Not only had the event lost access to the Black Rock Desert but its modest financial resources were wiped out—$300,000 worth of tickets had been sold in 1996, but much of that couldn't be found. With lax fiscal controls, there was no accounting for the missing money. The organizers, now deep in debt and facing insurance claims from one of the tent victims, peered over the brink at extinction.

In the face of all this adversity—perhaps *because* of the adversity that drew the core more tightly together—the show went on. Harvey invented a job for Will Roger, director of desert operations; Will took up residence in Gerlach to build bridges with the locals and figure out what to do next. Marian took over Mangrum's publicity job. Cleaning up Burning Man's unsavory image was mandatory. The local authorities had never liked the event, and now, given the deaths and chaos, they had ample opportunity to close it down, even were the organizers to find an alternative site.

The Temple of the Three Guys was dissolved by Law's withdrawal, and Black Rock City LLC was formed. Purists disdained it as evil capitalism incarnate, a HELCO look-alike, but the magnitude of exposure required a corporate buffer between the organizers and personal liability. "How do we expect to make Burning Man scale unless we form a more organized group?" was Harvey's rhetorical counter to dissenters. "If you're the possessor of real culture, there are great prospects if you're organized."[25] Black Rock City LLC incorporated with six owners, henceforth known as the Founders, each with an equal, nontransferable interest in Burning Man valued at $20,000—and, as conspiracy theorists were quick to note, rights to any profits.

In keeping with their devotion to experimentation and organic evolution, the original plan was to dissolve the LLC each year and create a new one. Fiscal exigency, however, quickly trumped idealistic intent. Continually re-inventing themselves hampered contracting with credit card bankers and renting office space. A governance protocol took shape. Whenever possible, decisions were to be made through consensus and community accord. A matter that "immediately affects the survival of Burning Man" would be the sole purview of the LLC owners. Collectively they called their deliberative body a Town Council, presided over by a director, the first of whom was Larry Harvey.

Now that the BLM had declared the Black Rock Desert off-limits, finding a new venue became critical. A piece of private property lay a few miles up the road from Black Rock City, owned by a local character named John Casey. It had a geyser on the grounds; during the 1996 Burn, he sold tickets to bathe in the hot springs. Many brush-ups with the BLM made Casey happy to irritate the Feds, and he offered his property. It was named after the aviation school that operated there in the 1930s, Fly Ranch.

The 1997 Burn at Fly Ranch was the great metamorphosis. Like a tadpole becoming a frog, Burning Man moved from one stage of development to the next. Out of a juvenile phase, with all the immature excesses that might have once been endearing but were now dangerous, into a preliminary version of maturity.

Unfettered creative exuberance remained central but tempered in Burning Man's "adult" iteration. "Our task this year," said Harvey on the eve of the event, "is to knit people into a group in which the overwhelming majority feel they're related to the others, and still have a part of the individual invested in the group."[26] In Harvey's idiom, Burning Man had launched its distinctive breed of civil society.

The most conspicuous difference from 1996 to '97 was physical. The contrast between the Black Rock playa and Fly was the difference between outer space and wilderness camping. "A broad and sweeping plain" was how the Burning Man newsletter described the new venue. "Cattle grazing on this flat plateau reduced its native grasses to a fine and intermittent haze which seems to hover." A veritable paradise compared to the desert. Complete with seven hot-spring ponds. "Glittering in the moonless night, our city will resemble some impossibly exotic seaport."

That Burn also saw the first proclamation of conscious ideology. The damage done by internal schism badly needed repairing but, even more important, Harvey seized an opening for a linguistic tactic that added sobriety to the intoxication of hyper-freedom. "Community" became the rallying cry. "If you look at the newsletter in 1997," he later observed, "it's one big propaganda sheet for community. That's when the litany began—community, community, community. People grabbed onto it."

Within the confined parameters of the ranch, the organizers could impose form on their city. They addressed the danger of reckless driving by banning the use of vehicles (with the exception of art cars) but that was just the start. Washoe County's festival permit came with stringent requirements. "We needed street signs, roads, signage, and an evacuation plan," recalls Will Roger, who was in charge of the build-out. "All these things we never had before. Every day was a new challenge. My creativity in solving the problems had to be very high and very precise. It brought out the best in me, dealing with regulations and problem solving."

The county, fearing hordes of Burners would become uncontrollable human locusts, demanded a master city plan. So Roger telephoned Rod Garrett: "I say 'I need a design' and hang up." A landscape architect who conveniently was also Roger's landlord in Oakland, Garrett had never before laid out anything comparable. "I haven't slept for two days," he said when he called back, "but here's my design." At the other end, Roger plugged his cell phone into a modem connected to a dot-matrix printer. The hastily drawn schema, a reaction to the disastrous consequences of the previous year's uncontrolled sprawl, became the crescent template that continues to define Black Rock City.

"Our goal was to express and abet a sense of communal belonging, and establish population densities that would lead to social interactions," Garrett wrote in his valedictory a decade later. "We were attempting to recreate some of the intimacy of our original camping circle, but on a much larger civic scale. Above all, this city needed to work without cars. It was vital that the flow of people and supplies in, out and within were unimpeded." To accomplish all these objectives, he came up with a conceptual breakthrough: arrange the city around a geographic center that functioned like the fixed point of a drawing compass. That point being the Man.[27]

A tribute to Garrett in *The New York Times* upon his death

in 2011 testified to his success. He was included in a short list of master planners—with Georges-Eugène Haussmann of Paris and Robert Moses of New York—who got to see their ideas take shape around them. "The Man is to Black Rock City what the Empire State Building is to Manhattan: a locating device and a reassuring beacon."[28]

Burning Man returned to public land the next year, having demonstrated to the Feds that the community could control itself. For all the problems with the BLM, they were easier to deal with than Washoe County, which regarded Burning Man as a cash cow to milk. Among the county's onerous demands was a requirement that fifty-five firefighters and four pumper trucks be on continual standby, with the organizers freighting the cost of $350,000. To ensure payment, sheriffs impounded all the revenues at the gate. "Robbed at gunpoint" is how another Founder describes the act. It left the enterprise stone broke.

But the die was cast and there would be no retreat. Larry Harvey's dream, which skeptics considered personally indulgent and needlessly grandiose, was in play. Attendees no longer thought of themselves as a band of lawless outliers; they were now members of that community. This identity encouraged people to take personal responsibility for the collective well-being. Black Rock City had a permanent framework, both architectural and social, to promote organic community. The foundations for a culture had been laid.

In response to the news that Burning Man was in big trouble following the great train robbery, Marian received word from a stranger in Austin, Texas, George Papp, offering to stage a fundraiser. "We never before had anybody do anything like that." The staff was just starting to use email, still a novelty, and

to publicize the event Marian assigned Austin@BurningMan
.com as the first external coordinates. Twenty-five people showed
up for Papp's event, one driving six hours through a rainstorm to
get there. Contributions totaled $500.

Emboldened by their unexpected success, the Texans staged
their own variation of Burning Man the next year. This would
be looked back upon as the original Regional. Countdown to
Armageddon they called it in year three. The progenitor stem
cell underwent its first mitosis, spawning a familial but unique
organism.

More such events followed. A network effect ensued, each
new Regional adding exponential value to the grid. By the time
Marian was speaking at the leadership conference, well over one
hundred events had occurred, some one-off and some ongoing.
They ranged from AfrikaBurn in South Africa to Fuego Austral
in Argentina to Dragon Burn in Shanghai to Crème Brûlée in
France. There were eighty-three in the United States. Even one
in cyberspace, the virtual Burn2.

Mainstream media caught on to what was happening out
there on the desert. *Nightline* ran a segment in 1997, likening
Burning Man to the dazzling new computer contraption, still
unfamiliar to most Americans. "Some people believe Burning
Man is the physical manifestation of the Internet," intoned
the suave anchorman, "a kaleidoscopic, no-holds-barred experiment." Then, with a we-all-know-better-than-to-take-
this-stuff-seriously twinkle in his eye, he concluded that Black
Rock City might be best understood as "a proto-Apocalyptic
neo-pagan freak-fest."[29] The seeds of what continues to be the
prevailing media narrative were sewn, the endless retelling of
the tale of bacchanalia in the badlands.

The buzz in tech circles, however, was considerably different. Convinced that they were spearheading a movement where "information wants to be free," the digerati recognized a kindred spirit. Burning Man, like them, was all about the democratization of content without censors or gatekeepers. Black Rock City became the venue of choice to try out wild ideas, a proving ground for cutting-edge prototypes (Google Maps would beta-test here). The Founders weren't techies but they were breathing the same heady Bay Area vapors. A symbiotic partnership took shape.

Wired magazine, the bible of the tech revolution, put Burning Man on its cover in November 1996, proclaiming the event "the new American holiday." Prior to the appearance of the piece, Larry absent-mindedly mentioned to Marian that a reporter for some magazine had interviewed him. He thought it was no big deal. What magazine? *Wired.* "Everything is going to change," she exclaimed, positively delirious. Burning Man was about to be endorsed as a high-status place to see and be seen. At first sight of the issue, she teared up.

Marian's importance within the leadership, "the resident adult" one journalist called her, grew. "A no-nonsense blond with a voice that sounds like it's barking orders out of a walkie-talkie" was another description.[30] In a different time and place, she would have been well cast as the chair of the Junior League or perhaps the feisty frontier sweetheart in a cowboy movie. At Burning Man, she was the rising star of back office operations.

Burning Man's historical arc might look inevitable in retrospect, but it was periodically within a match-strike of blowing up.

The Great Art Controversy of 2004 was one such moment.

In its public posture, Burning Man was being positioned as a colossal art show. By then the budget exceeded seven million dollars, with a major line item of nearly three hundred thousand dollars going to support art projects. A handful of pieces got grants, awarded by a committee headed by Larry Harvey. The system worked tolerably well despite mounting complaints from the iconoclastic, "kick-ass" old guard that found selections increasingly boring. Matters came to a head when there was widespread discontent, even rumblings of rebellion, that the art scene not only had deteriorated but was rigged on behalf of Harvey's favorites.

Protest arose when a group of high-profile Burn artists, the rock stars of the playa, proposed creating a second funding body to be financed by 10 percent of the ticket sales. Petitions circulated, accusations exchanged, debate swirled. There was even mention of staging an alternative event. "This year we have 1,000 people," said one of the dissidents, "next year we'll have 10,000 people behind us." The LLC launched a counterattack, conceding that the projects of 2004 were less daring but dismissing this as simply a bad year rather than a conspiracy. The tide of opinion turned when a celebrity collective, The Flaming Lotus Girls, opted to remain inside the LLC tent.

The art installations in 2005 proved a colossal hit. More projects were funded at higher levels than before. Harvey denied that the reason more money was funneled to artists that year was to quell dissent, but Marian dismisses his claim. "If Larry tells you it wasn't, he's full of shit." *Angel of the Apocalypse* was a star of the show, a sculpted twenty-foot-tall bird with gigantic wings lit at night by liquid flame. Its creators: The Flaming Lotus Girls.[31]

Consensus governance, another foundational pillar of the culture, was also tested. The consequences of failure would have been devastating to Burning Man's mystique. In this instance, the controversy involved music. Dance camps, with their thundering, pulsating acoustics pumped out by monster speakers, started showing up in the 1990s. EDM was not universally welcome. In principle and practice, ravers fit into the Burn ethos, but their reputation for illegal drugs and reckless behavior posed a serious political liability. Local officials wouldn't permit them. "The cops told us," recalls Harley Dubois, "that if they hear the word rave, 'we're shutting you down.'"

Moreover, it was no secret that the Burning Man establishment . . . *hated* . . . trance music. "The top dogs just weren't into it," remembered a rave leader dismissive of the old guard. "They wanted to create a counterculture that they thought was *the* counterculture."[32] The younger ravers, now inside-the-fence physically and metaphorically, refused to go away. A generational divide festered: metalheads vs. ravers, Led Zeppelin vs. star DJs. An accord was cobbled together: a one-time rave for eight hours, innocuously called the Community Dance to keep the police at bay. It was pretty much a dud (cold weather, bad vibes) but the precedent had been set. A major crisis simmered. Lots of people wanted electronic dance music in the city, lots didn't.

Harley Dubois, who had come up with the idea of placing big theme camps in strategic spots to delineate plazas and an esplanade, was now Black Rock City's de facto zoning czar. Resolving the crisis fell to her. The way she proceeded became the culture's prototype for civic decision-making.

Her first effort split the town into "Loud" and "Quiet"

sides, logical but unworkable in practice. Even hard-core party types needed sleep occasionally, free of disturbance from their neighbors. And because of the crescent shape of the street grid, sound still noisily reverberated into the quiet zones. The lasting solution, several years in the making, came out of extensive deliberations among a classically diverse Burning Man constituency: ravers, circle drummers, sound engineers, professional club bouncers, and Michael Mikel's volunteer peacemakers, the Black Rock Rangers. The answer: locate the loud music at the two edges of the city facing the huge speakers outward, toward the distant playa. "We created zones at either end of the semi-circle for officially named Large Scale Sound Art camps, making no value judgments about the musical quality," recalls Dubois. It worked.

The solution might seem obvious now, at the time it wasn't. Nor was getting everybody to agree easy. But Harley's patient process of reaching accord step by step, taking as long as was necessary, would henceforth be embedded in the culture. Decisions wouldn't be reached through compromise, where 51 percent trumps the 49 percent who are obliged to obey, but rather via a consensus model in which any stakeholder can exercise veto power until there is a universally acceptable resolution.

The sound art controversy looms large in Burning Man's institutional memory as the critical moment when its governing model was validated. It reaffirmed the Founders' commitment to unraveling conflict without compromising values. Sure, solving the noise hubbub was important but more important was enhancing the coherence of the community. If Burning Man culture were to be sustainable, which for Larry was pretty much the whole ball game, it had to be able to bend without breaking apart. The process Harley pursued—open, flexible, experimental—henceforth became Burning Man's gold standard of governance.

Two decades later at the Global Leadership Conference in San Francisco, Marian was reassuring people that nothing would change under the Project. Burning Man's central office would continue to differ from a conventional headquarters. It didn't issue directives but rather convened conversations. Its job was to bring people together, hear diverse voices, lay out options. Not issue orders. An undercurrent of tension between headquarters and aspects of the community inevitably comes with the turf, but Marian's message was that the Project would continue to be the guardian of the cultural DNA, not author of its code.

Still, in its empowered organizational mode, the intention of the Project was to extend Burning Man's influence. A professional fundraiser was now on board. Previously autonomous outreach programs—the Black Rock Arts Foundation, Burners Without Borders—had been brought in-house, consolidated into an Office of Art and Civic Engagement. As Marian strides back and forth in the ballroom, her excitement bubbling over, she is in effect talking about leveraging Burning Man's impact.

She finishes the session with an emotional crescendo, recounting the moment she came to believe, *really* believe, that Burning Man's destiny was to meaningfully shape the world. It happened in Israel, where the regional Midburn ("Midbar" in Hebrew) was a stunning success. It was started by Nir Adan, a third-generation Israeli who, while visiting cousins in New York, ended up at Burning Man "super accidentally with no idea where I was going." He returned thirteen consecutive years. Back in Israel, he'd hang out occasionally with a small group of

Burners but with no thought about staging an event. Then, to celebrate his mother's birthday, he decided to host a Mama Burn on the beach (which was the only place you could have a public gathering without a permit).

Sixty invitations went out, three hundred people showed up. The next year, attendance topped fifteen hundred. "We decided we are going to make an event that makes a community or a community that makes an event." Midburn launched in 2013; within two years, eight thousand people came. In a region where continuous political tension can breed suffocating conformity, "Here's this new bubble that offers you freedom of choice." Dozens of Arab Israelis attend. A contingent from Dubai has come, very quietly and under the radar. A pan–Middle East Regional is even being discussed, an instrument of reconciliation disguised as a festival.

This was the backdrop to Marian's story. She'd been invited to attend a wedding of Israeli Burners. The venue was a mountaintop overlooking the West Bank. Amid the joyfulness of the festivities, she gazed over the contested landscape below, where generations of violence and suffering have unfolded. In a flash of intuition, she was filled with a certainty appropriate in this biblical setting. She just *knew* that Burning Man had much to offer warring people, that it offered a way to lessen enmity. Its fundamental cultural truth—that surrendering self-imposed personal restraints about "reality" enables human creativity to redefine social possibility—was a missing link to peace.

Burning Man could, indeed, change the world.

3

"A New Moment to Gather"

What would it mean to have a year-round location beyond the playa? What if we had a place to experiment with and apply the Ten Principles 365 days a year?

"We Bought Fly Ranch,"
The Burning Man Journal, June 10, 2016

Midway through the 2016 Burn, six months after the Global Leadership Conference, a goodly crowd of people ducked out of the open-hearth midday heat, shed their shoes, and assembled beneath the cool shade of a big-top tent. By now the news was out that the Burning Man Project had acquired Fly Ranch;

it was the buzz of the playa. Here, in what was known as the Da Vinci Dome, a public conversation about the property's future was about to begin.

For the next hour and a half, in the same space where throughout the week would be sessions in collective breath ceremony or physiological synchronization or healing the rift between one's masculine and feminine selves, several dozen community members came to hear what the Project had in mind for the thirty-eight hundred acres it had bought some fifteen minutes up the road from Black Rock City.

The purchase was the culmination of decades of dreaming and trying to cobble together the financing. After all that effort and time, the leadership was in no hurry to act precipitously. This gathering was an opening round in discussions about the future of the property. The process of deciding, in keeping with governance by consensus, would last as long as it took. Only one deadline had been set: The next two years would be devoted exclusively to "listening to the land and listening to the community."

The art theme in 2016 might not have been entirely self-serving, but Da Vinci's Workshop was remarkably attuned to the Fly Ranch announcement.

Harvey was effectively removed from operations under the Project. When it was being assembled, he half-heartedly suggested he should be the Project CEO, but it was obvious to all that his talents weren't managerial. Marian came up with the appropriate position, chief philosophical officer, in charge of something to be called the Philosophical Center. He became by title Burning Man's resident thinker and guardian of its institutional consciousness. Which he had always been, but now it was an official job description. Even in this role, however, he continued to exercise considerable power. The covenants of the Project specified that Harvey alone got to choose the annual art theme and, in conjunction with that, the design of the Man. Each year's Burn became, in effect, his personal art project.

Da Vinci's Workshop was his idea. And a calculated ploy. He told me as much before the event, as we chitchatted on the roof of the headquarters building. It's situated on Alabama Street, among warehouses and workshops converted into contemporary office lofts, where it had moved from a considerably seedier Mission District neighborhood. The Project occupies the two top stories of its building; the open floor plan is airy and bright with conversational areas to the sides. Art lines the walls, curated to tell the history of Burning Man. The views from the roof are panoramic. Harvey often conducted appointments there so he could smoke.

The Fly Ranch deal had been announced a few weeks before we met. Harvey was understandably excited and eager to start shaping a narrative to engender support for the scale of the enterprise he had in mind. Indeed, he explained to me, his intention with this year's theme was to start nudging the community to see itself in a new light. As patrons, modern-day Medicis, underwriting a second Renaissance.

Not that he was equating twenty-first-century Black Rock City to sixteenth-century Florence, but Harvey was quick to point out similarities. Just like the Italians, Burners were responding to a turning point in history with renewed commitment to humanism and inspired art to back it up. "Think big," was Harvey's message, "you too have the opportunity to change civilization." The talents embodied within the community could

make Burning Man the vanguard of humanity's next great
thrust forward, the epicenter of a cultural revitalization. The talent was there, the will, the freedom. Only one thing was missing, the essential catalyst. Money!

He was convinced this was just a temporary obstacle; the
money was out there. The founding generation of tech billion-
aires had Burning Man in their blood, as did plenty of other
people of means. Harvey's subtle strategy to arouse the one per-
centers in his constituency was to get them to think of themselves as the financiers of civilization's next advance. Which
would happen through grand expressions of Burning Man culture. At Fly Ranch.

Given the prevailing cultural catechism, however, any talk
about money within the Burner community is immediately suspect. Black Rock City by design is a no-cash zone. You can't buy
anything on playa, except ice (mandated by the state of Nevada)
and coffee (a tradition from the first days on the desert when a
pot of coffee was always simmering at center camp). There're no
souvenirs for sale, no convenience stores.

Nor is this a barter economy. When something changes
hands, it's free. No quid pro quo. What *is* available—from cold
beer to artisan pancakes to full-body massages to handmade
jewelry to bicycle repairs—is gifted. Gifting is the fundamental
currency here. Social capital is the medium of exchange.

In Burning Man's cosmology, removing cash from interactions makes them more spontaneous, more likely to engender
empathy and gratitude, more likely to elicit human emotions.
Money sparks class resentment; gifting fosters community. In
the outside world, the cash nexus of buying and selling con-
taminates those selfsame human sentiments. The subtlety of the

argument isn't lost out on playa, where the zeal for offering gifts is ardent. So for Harvey to even obliquely introduce money into the conversation, to surreptitiously hint that crossing the next great creative threshold depended on financial patronage, bordered on heresy.

The rise of plug-and-play camps had brought the subject of money to the fore. These are for-profit pop-up luxury operations offering full services. No sweat equity here. "Campers" reside in swank RVs, eat gourmet food brought in and prepared by metro-chic chefs, and have every need attended to by a concierge crew masquerading as Sherpas. Unlike the welcoming spirit within the rest of Black Rock City, these camps are typically sealed off, the vans drawn into a tight circle with a monitor at the entrance. Veritable gated communities. The price for such exclusivity can be as much as $40,000. De-commodification be damned!

To the rank and file, plug-and-play camps are kryptonite; they parade entitlement. They have drawn considerable criticism. Even vigilante vandalism. The media has picked up on this contradiction; a newish meme is that Burning Man surrendered its old-school rigor and self-reliance to the privileged class.

The Project leadership has sought to thread the needle on this one, defending the presence of such camps by virtue of radical inclusion, while putting up obstacles that make it more difficult for them to operate. They've been constrained but not censured. "What I think these camps are really guilty of is being gauche," said Harvey, seeking to calm the waters. "This is not so much about morals, it is more about manners, and we're convinced bad manners can be mended." The thinking at headquarters was that accepting turnkey camps was not only a democratic gesture but, perhaps more important, it exposed a demographic of the rich and powerful to the Burning Man virus. To pull off the Project's big dreams, people of affluence and influence—Harvey's patrons—would be invaluable.

Harvey, wizard of mass psychology, conceived this year's theme as a preemptive strike to reframe the bias of those who pooh-poohed money. Da Vinci's Workshop was an enticement, a kind of luscious bait that hooked the crowd with spectacle for the sake of changing their attitude. As we basked in the uncommon spring sunshine of San Francisco up there on the roof, he spelled out his plan. "The community has this quaint notion going around that money corrupts. They see Black Rock City as a utopia that is money-free." His tone, not quite scowling but clearly exasperated, showed his impatience for such naiveté. He rattled off a list of things that cost lots of money without which Burning Man couldn't happen: permits, construction equipment, all those porta-potties (the single biggest line item on the operating budget). Not to mention some hundred-odd full-time staff members. Some of the more vocal ideologues were living in delusional wonderland. "They think they're getting absolution from consumer culture in an eight-day sojourn," Harvey said, "but we're saying they're looking through the wrong end of the telescope."

By using the Renaissance theme to advocate patronage, he was out to reconfigure the wisdom of his crowd. Just like a handful of Florentine families underwrote great churches and libraries, so had a handful of rich Burners—including a cofounder of Airbnb, an early investor in Twitter and Instagram, a Facebook vice president, the guy who started out as a street performer in Quebec and ended up creating Cirque du Soleil—endowed the community with Fly Ranch. To Harvey's thinking, this marked a new phase in the evolution of the community. He envisioned contributions, lesser in size but comparable in commitment, from thousands and thousands of Burners.

The Da Vinci Dome, where the opening Fly Ranch dialogue was underway, was part of Red Lightning camp. Over twelve hundred camps spread through Black Rock City, ranging in size from hundreds of people like this one to a handful. Red Lightning, by virtue of its location on the Esplanade, was something special, at the innermost curve in Rod Garrett's crescent. To be placed so conspicuously was a mark of importance. Behind it, for maybe half a mile, streets unfold in alphabetical order, the names changed in keeping with each year's theme. In 2016 they were Arno, Botticelli, and Cosimo through to Knowledge and Lorenzo.

"The commitment of the camp is to build a culture of ceremony and transformation" is how founder Brad Nye explains Red Lightning's mission to me. "To create for humanity a New Story that is waiting to be lived." Translation: Promote a contemporary mythology more compelling than old-school ideologies. "People are living in a meme that isn't allowing them to live in their highest potential. Our role is to create systems that help the collective wake-up."

As with so much about Burning Man, technology and engineering render plausible propositions that otherwise would be garbled madness. Structures that just *couldn't* be, that are too impossible to exist here in the middle of the desert, appear everywhere. In Red Lightning's case, it's the Dome, a spacious wonder rising to a height of 60 feet with audio and visual capabilities designed to alter perception in what can be best described as 360-degree immersive group reality.

The folks meeting in the dome that afternoon didn't need sensory prodding; they were already primed. Fly Ranch was going to afford Burning Man a year-round presence. It would be the physical fulcrum to leverage the culture outward. The question was what to make of this property—what should this permanent presence be and do?

Ideas bubbled forth: an art park; a direct crisis intervention facility to support, say, refugees; experimental centers for child development, architectural experimentation, healing pain; a human potential retreat like the Esalen Institute. A self-described student of economics observed that, since capitalism is about value extraction, Fly Ranch could model a different scheme, "value production through participatory commerce." The slate was blank.

Will Roger, one of the founders, rose to offer a benediction as the session ended.

Roger had earlier told me a bit about himself. He grew up in a suburb of Buffalo, New York (as did I, Amherst to his Kenmore). With a degree in analytical chemistry and a master's in photography, he came to the Bay Area to set up a studio in Oakland. "I've had my trials and tribulations," he admits. "My problems with addictions and alcoholism. What happened to me was I bottomed out, I hit that point where you make a choice if you want to live or die."

Nobody has a keener love of this Nevada land. His permanent residence is now nearby Gerlach; he and his fellow Founder wife, Crimson Rose, have a home with a hot-tub spa on two acres surrounded by some fifty apricot, peach, and apple trees. He only leaves to commute to San Francisco for board meetings, where he served as the Project's first chairman. "I'm very good at running meetings." Ask about the Black Rock Desert and he'll speak with rapture about the rainbow hues of the mountains or the annual campout he leads deep into the playa to watch the Perseid meteor shower.

As the keeper of the faith all these years, Will Roger's pride bubbled exuberantly as he spoke to the folks at the dome about Fly Ranch. He sketched out the arc of history that led them to this moment, starting with the Cacophonists. "They embodied the young, fiery, transformational energy of the late twentieth century." Now, however, different times called for different responses.

The agenda had changed. "It's not about burning down a magic city but rather building it up." A genuinely modest man with a taste for cigars, Roger in his gentle manner belies the full resonance of his personal power. "This is," he proclaimed, underscoring how significant this meeting would look in retrospect, "a new moment to gather."

II

CHAOS AND CONTROL

The Burning Man Experience

4

Da Vinci's Workshop

One Man's Burn

You don't always get the Burn you want, but you
always get the Burn you need.

Playa aphorism

From the outside looking in, few would have noticed much difference in the 2016 Burn from earlier ones, Will Roger's invoca-
tion of "a new moment to gather" to the contrary.

When someone who's never been asks what Burning Man
is like, inevitably you shrug. "Impossible to explain" is the universal default. Da Vinci's Workshop was no exception. Why so
hard to put in words? "It is an event rich in parody and paradox,"
answer a couple of scholars. "Just as one might seize on some

aspect of the event that might promise to represent its 'essence,' another aspect comes to the fore and pulls the proverbial rug out from under any such temptation."[1]

For starters, there's the problem of translating size into comprehensible terms. Everything—the spread of Black Rock City, the size of the art, the kaleidoscope of dazzling lights at night, the concentrated mass of people—is vaster than anything you've got words to describe. Scale works the other way, too. Tiny pleasures, the simplest things that would otherwise be unremarkable—a juicy orange, clean socks, a shower—are prized and appreciated. Then there's the unique setting. Dust is everywhere. Caked on your skin in an alkali patina impossible to wash away, in the folds of your shirt and crevices of your backpack. It's thick enough to burn your eyes during the daily afternoon windstorm when the temperature starts falling (one learns to always keep protective goggles at hand). And how do you make gifting comprehensible? From cappuccino and croissants in an impromptu café to an entire thrift store's worth of wardrobes with a seamstress on hand for custom tailoring. Or how to describe the ultimate phenomenon, when whatever you need, from a soul mate to a Band-Aid, mysteriously appears on cue? Those endless occasions of serendipity known simply as "playa magic."

Black Rock City is a 24-7 sensory overload. Bizarre get-ups, turbo-charged erotic pheromones in the air, dazzling LED lights, mutant vehicles morphed into unicorns and butterflies and nineteenth-century frigates, wild temperature swings from boiling hot to freezing cold, the aroma of bacon, the ecstasy of a hammock in the shade or an unsoiled porta-potty. Even the sweet surprise of a passionate kiss from a beautiful stranger as she walks past you to forever disappear.

Then there's the satisfaction that comes from the sheer act of survival. Of self-reliance. Having to solve life's little technical glitches yourself is very much woven into the texture of experience. I'm forced to become more resilient, more inventive, more competent. All of which feels . . . empowering! Nor am I alone in this. Our small camp, sixteen people in 2016, had to plan in advance to lay in a week's worth of food, kitchen equipment, and an electrical generator with enough gas to keep it running. Something as basic as water can't be taken for granted. Delivery of potable water has to be arranged (some tanker trucks come from as far away as San Jose), gray wastewater hauled away (a camp that leaves litter risks being banished).

I was brought unwillingly to my first Burn in 2013. Cait, my wife of twenty-some years, announced her plans to attend, whether or not I accompanied her. All I knew about Burning Man at that point was its reputation for sexual licentiousness. Damn right I was going to accompany her! But dragged kicking and screaming. As the moment to shove off from Reno neared, I grew near-catatonic. That's how anxious I was. Afterward, back in D.C., friends asked how I did. They were expecting the worst. Now, I've traveled many places, including Madagascar and Myanmar. "Best trip I've ever taken," was my answer. "Can't explain it but the whole thing was great." Which pissed off Cait no end, since, as she often complains, "I was the one who wanted to go and you ended up having more fun."

It was the social mix, mingling with divergent types with time enough to talk, that I found such fun. "No one constituency has taken over," writes John Curley, onetime news editor of the *San Francisco Chronicle* who now blogs for the Project. Of course, as advertised, there's a big party scene: ravers on designer drugs, hordes of frat boys and glitzy sparkle ponies looking to hook up, blockbuster sound camps funded by Russian oligarchs with imported German DJs. And dedicated sex venues in all flavors, including the fabled Orgy Dome ("consensual play space for couples and moresomes").

An occasional celebrity shows up; you know who's here by their social media. It was Paris Hilton in 2016. Along with Heidi Klum, Katy Perry (her braids interwoven with rainbow yarn), Karlie Kloss, and Leonardo DiCaprio's love interest at the time. Purported sightings of tech heavyweights, the likes of Sergey Brin or Elon Musk, are commonplace. One year, according to legend, Mark Zuckerberg arrived by helicopter. Politicos and government officials have been known to come, usually incognito and well concealed. But representatives from either end of the ideological spectrum, Grover Norquist and Dennis Kucinich, both made public attendances this year.

Among the 75,000 are also solid middle-class types. Doctors and lawyers, working stiffs with union cards, families with children. There's a whole Anonymous Village to provide safe shelter for 12-Steppers. For every workshop in the ultimate orgasm there's another in meditation. "It's not ALL about any one thing at all," concludes John Curley. "Which keeps the event vibrant and interesting, even while you judge some things you discover there ridiculous and abhorrent."

Burning Man is decidedly, adamantly, aggressively *not* an entertainment venue. The misguided sometimes mistake it for Coachella, the California music extravaganza, but here there are no marquee acts. Indeed, the Project loathes to call it a "festival." Nobody spectates. Everybody's part of the show. The annual guidebook for Da Vinci's Workshop ran well over one hundred pages, offerings from the community to the community. Thumb through the book and you see titles like Kamikaze Karaoke, Mindful Wine Tasting, History of Bluegrass Music in Words and Song, Mask Making, Belly Dance Basics, Bee Keeping, Advanced Paper Airplane Folding, Ukulele Jam Session, Negotiating for Women, Milk & Honey Shabbat Service, Symposia on the Neuroscience of the Mind, even a Cotillion Ball in Illuminated Hoop Skirts. And all kinds of sex stuff. Intros to bondage and polyamory, naked shot parties, masturbation circles.

Da Vinci's Workshop was Cait's and my fourth straight Burn, the last three accompanied by our son, William, who started going at eleven years old. We are, according to stereotype, most improbable Burners. I am a white-haired grandfather whose résumé includes Peace Corps service in the Ivory Coast, a graduate degree from Yale, a few years as a college professor, and then a career in journalism with some blue-chip gigs. Cait is even *more* respectable, an attorney with a Harvard doctorate and a high-level job in the federal government.

William, as a "Baby Burner" is, admittedly, pretty special. A bronze-skinned lad descended from ancient Mayans, his natal mother emigrated from the western Guatemalan highlands, crossed the Sonoran Desert on foot, and ended up in the chicken processing plants of Delaware (where, a while back, the companies busted the unions and replaced the workforce with underpaid immigrants). We adopted Will at birth, with Cait in the hospital delivery room watching him enter the world with an outstretched fist that we've always regarded as a harbinger of his prowess.

The point of all this is to underscore that, contrary to what you've seen on television, we are much like thousands of others in Black Rock City. Not everybody is a freak or a hippie. Many are responsible parents, monogamous couples, untattooed and, with the exception of an occasional hit of weed, drug-free. Dare I say it, bourgeois in lifestyle if not temperament.

What I didn't know when I arrived the first time but quickly discovered is that I am a natural for Burning Man. Not in the obvious ways: I don't wear costumes or revere Black Rock City as my spiritual home. Instead, I simply . . . hang out. I've come to love the desert atmosphere, the sweetness of early mornings and the cooling respite when the sun begins to set. I always find something of interest as I ride around on my bicycle, the pre-ferred mode of transportation. And the conversations! I've been nowhere else where they are so consistently satisfying, often with total strangers.

I still have to fight down pre-game jitters when the moment to commit comes each year, but my love of the whole scene had only kept growing by the time of Da Vinci's Workshop. Which was when I became Crow. Playa names show up when they're ready. Mine appeared the first morning I was on playa, during Build Week, three days before the official opening. I had arrived the previous night in the pitch-black wee hours. When I awoke with the light of day, I shrieked with joy. Far off, somebody answered with a loud set of caws. "Sounds like crows," my pal observed. And thus was I christened.

As Crow, I encountered peak experience after peak ex-perience after peak experience throughout Da Vinci's Work-shop. With little dead time in between. Exhaustion, perhaps, but never boredom. "Mindless bliss" might be your local guru's term. A spacey reverie, owing as much to overbaked brain cells as transcendental introspection. Not that there aren't moments of misery: anger and frustration, irritability, sleep deprivation, moody crankiness. Being bummed out contributes to the whole gestalt. Sharp mood swings are part of the drill. Taken together, these acute feelings of good and bad start to feel granular. Time becomes physical. The mélange of the week fuses into interstitial matter.

For whatever reason—the setting, the self-selection of those who attend, the aforementioned Zen zone-outs—the default at Burning Man is to embrace who and what shows up as part of the same corporeal body. Colleagues in a colossal circus of trustworthiness and compassion, empathy and generosity of spirit. As an ensemble, the citizenry constitutes a make-believe order. The Brotherhood and Sisterhood of Fanciful Possibility.

Most of the players not only drink the Kool-Aid, they crave it. Punch-drunk on existential Red Bull. To a fault, sometimes. A lot of the high spirits are drug induced, to be sure. Psychotropics consumed at Burning Man in high doses "jacks consciousness out of its usual ruts" and seems to "catalyze awesome experiences of cosmic fusion."[2]

This intoxication with possibility comes back full circle to the seeds sown by the Founders. Simple-minded as it sounds, like the group cheer at the end of a motivational weekend, there's no getting away from the fundamental truth of this place: people operate on a different plane than in their ordinary lives. They are consistently revealing parts of themselves usually kept hidden, stripped of pretext, raw and vulnerable, courageous. Or, at the other end, selfless and compassionate.

Make no mistake, though, danger also lurks. "The devil's pinball machine" is one habitué's description of Burning Man;

another calls it "a place where exceptionally self-centered behavior is exalted and glorified."[3] Somebody's version of paradise can be another's definition of hell. As evidence of this two-edged sword, I submit this description of the morning after a night of polyamorous playa libido. "Group sex, while totally awesome at the time, can turn into a lot of long, really boring conversations the next day where everybody needs 'to process.'"[4]

Meltdowns are part of the landscape. Like my daughter's. Amelia accompanied us in 2014, a D.C. millennial in her late twenties. Vibrant, smart, talented. Well educated. Employed. Hordes of friends. She couldn't have been more excited, she just *knew* the Burn would be fantastic! Which proved to be her undoing. Expectations, as one playa mantra goes, are your second biggest enemy after the desert. "You just gotta learn to let shit slide."

Bad karma plagued Amelia: cliquish campmates, thwarted plans, a stolen bike. Downward-spinning spirals. On her home turf, she projects confidence but here amid so much disorientation she was consumed by uncertainty. What might otherwise have been a trivial disappointment sent her over the edge. A skilled chef, Amelia randomly met a fellow foodie who offered an invite to cook together at the women's camp. Amelia showed up at the appointed hour but didn't spot her friend. She paced back and forth on the outer edge of the camp, not daring to enter. Stunned by her unexpected timidity, she broke down in tears. Recollections of past setbacks, of past failures, bubbled up over the next days. Amelia was not the first nor last person at Burning Man to become overwhelmed by deep-seated self-doubt.

In her case, there's a happy ending. Chastened by this

wake-up call, she came out the other end stronger and healthier. She's made major, positive life choices. "You don't always get the Burn you want," goes an oft-repeated playa maxim, "but you always get the Burn you need."

The Burn I must have needed at Da Vinci's Workshop ended up being more subdued than exuberant. Perhaps because I was on playa for thirteen days and twelve nights, arriving early to build out our camp and staying late to break it down, it had the steady rhythm of a marathon. One particularly wonderful "holy shit!" moment defined that year's enchantment: my jam session with the Hollywood guys.

A chain of connections led me to David Silverman. Original animator of *The Simpsons*, director of *The Simpsons Movie*, and co-director of *Monsters, Inc.* I sought out David to interview him for this book. As we drank vodka-infused slushies, I asked how he has been affected by Burning Man. "It helped me become the musician I am," he answered, and spun out the tale.

In high school, he took up the tuba. "The size intrigues people, they like hearing the rhythm." He started bringing it to Burning Man, figuring out how to wick the horn and shoot flames out of the bell. Hence his name, Tubatron! Which eventually led him onstage for the grand finale ensemble of *The Simpsons Take the Hollywood Bowl* concert in Los Angeles, blowing fire from his tuba. "All my worlds collided on that stage!"

As if on cue, at this point one of his campmates wanders over. Michael Giacchino joins us, a celebrated film composer who's won an Academy Award for *Up* and scored *Star Trek* and *Ratatouille*. We continue to discuss music when, impulsively, David asks if I'd like to hear him play tuba. Well, of course! He goes off to fetch it, and Michael gets his drum kit (which he confides he's been playing for only a year). Minutes later, we've got a gig going out in the street. David on tuba, Michael on drums, and me—uninhibited after a second one of those midday slurpies—scatting as the front man. A crowd gathers. Michael produces a goody bag of percussion instruments and passes around tambourines, rattles, triangles. Voilà. Burning Man culture at its essence, a pop-up participatory jam with two (anonymous) stars.

Music kept showing up in 2016. I did karaoke for the first time, singing the song that got me through my divorce years earlier, "The Heart of the Matter," with a conviction worthy of Don Henley. I was randomly gifted a dime-store kazoo, carried it around in my backpack, and, perhaps inspired by Tubatron, launched into impromptu riffs all week. The Playa Jazz Café, another "gifted" venue that offers drinks while folks await their turn to jam with the house band, became my regular late-night haunt; one evening the keyboard guy hoisted himself upon a rafter in a moment of unbridled gusto to pound the ivory with his toes.

"Sonic Runway," an art installation deep out in the far reaches, was my Big Time Mind Blow. Thirty-two arches strung with LED lights formed a thousand-foot-long passageway wide enough to ride through two abreast on bicycles. At the front end of the column were audio sensors, microphones to pick up acoustic vibrations from ambient sound (usually resounding music from tricked-out art cars). Which would then trigger complex configurations rippling down the corridor, algorithms of color passing from arc to arc like breaking surf on the beach. I rode through the Sonic Runway during Build Week, bedazzled by the flashing lights but ignorant of what was happening.

The Mind Blow occurred during my second visit to the Sonic Runway, when Cait and I brought Will out to see the art scene at night. By then, I had learned how the piece worked. The patterns rippled down the chute, from beginning to end in less than a second. Which translates into 768 miles per hour. The speed of sound. Standing outside as they walked through (the crowds too thick for bicycling), I now saw clearly how separate patterns maintained their integrity flowing down the channel. This wasn't amusement park razzle-dazzle, flashing random lights, as I had thought the first time. Rather, these were light quanta traveling at the speed of sound. *I was seeing the speed of sound!* Mull that over again. A cohort of inspired sculptors, computer immortals, and animation masters had . . . gifted the community . . . a super-fun interactive installation that enabled us to *witness* the speed of sound.

Fatigue caught up with me as the week went on. That first shriek of the crow felt like ancient history. My overloaded senses dulled. This is normal, I was prepared. Virgins typically aren't. Veterans know to take it slow in the early days, to conserve their strength for the inevitable exhaustion near the end. On Saturday, as I was in danger of succumbing to weariness, a palpable buzz in the air brought me back to life. This was Burn Night.

In keeping with Harvey's ambition that Da Vinci's Workshop be a cultural awakening, he conceived this year's Man in epic terms. It was modeled on Leonardo's Vitruvian Man, the iconic sketch of a figure whose outstretched arms simultaneously inscribe both a circle and a square. The archetypal emblem of the Renaissance.

Harvey called his homage Turning Man. It was designed to raise and lower its arms through an elaborate system of gears and pulleys. The driving force for this dynamo was to come from sixteen volunteers tugging together to rotate the effigy, a full 360-degree revolution calibrated to take an hour to execute. Upside down and back again, "as if it formed the axle and spokes of an enormous spinning wheel." Never had interactivity, the touchstone of Burning Man's aesthetic, been attempted with such ambition. And then, so the plans went, just before ignition would come a dramatic crescendo. Harvey would sit at the controls of a motor and spin the effigy around and around and around.

None of this happened.

Making a 43-foot-tall, 20,000-pound giant rotate pushed the envelope a few nudges too far. When the gates officially opened at midnight Sunday, Turning Man remained unfinished. He stayed headless for several more days. Rumors flew. Finally, on Tuesday, the building crew gave up, unable to fix the drive mechanism (when the figure was mounted, excess torque sheared connection bolts in the transmission system). Turning Man, head upright and locked in place, was immobile. Which, in truth, little mattered to the gathered multitudes, few of whom knew what had been planned.

As the geographical and emotional center of Black Rock City, a towering point of reference as well as a psychic landmark, the Man is treasured. The first time I glimpse him each year, it's with the warmth of spotting an old friend. That his existence is brief and bittersweet, that his destiny is to be sacrificed, makes him all the more endearing. The cause to which he is martyring himself remains unstated: Material transcendence? Dada anti-art? Primal fire worship? The meaning of the Man resides in the mind of the beholder.

His burning this year almost didn't happen. The artists, construction managers, and pyro crew had done their thing in San Francisco, figuring out how to weaken and compromise the structure to collapse in a controlled burn. What engineering couldn't calculate in advance was the weather, which this Saturday was bad. The set had been readied the night before, electrical equipment (for the aborted rotation) stripped out lest the smoke be poisonous, but all day dust storms gusted too fiercely to conduct a safe burn. Then, at the appointed hour, the winds abated. Playa magic!

If all roads once led to Rome, on this night they led to the Man. Thousands, biblical hosts, scurried to the site to participate in the Ritual of Rituals as the chorus of observers. The pageantry began. Crimson Rose is responsible for its design. The Procession of the Ceremonial Flame, carrying the sacred fire "extracted from the flame of the sun" via magnifying glass and tended all week in the cauldron El Diablo, began the half-mile march from Center Camp, preceded by white-robed bearers of ten-foot torches. At the Great Circle, the perimeter surrounding the Man, groups of dancers—the Fire Conclave—executed elaborate choreography while holding flaming staffs. The dark rumblings of the Ambient Drummers resonated in the background. All was ready for Crimson Rose to ignite Turning Man.

Within the Founders mix, it is Crimson (not her birth name, which very few know) who is most responsible for Burning Man's well-honed theatricality. She moved from her home in Southern California to San Francisco in the 1970s, earning her living as a fine arts model ("I was a professional naked person"). Her passion was dancing; she had been introduced to fire dancing well before Burning Man. The tradition traces back to the Maori people in New Zealand, who used ignited balls on a string to train for war. San Francisco caught on to fire dancing in the 1980s, where Crimson practiced a genre called fleshing. It, too, derived from tribal traditions; she was dipping torches into a bowl of fire "and laying that on my body." She plugged into Burning Man in 1990, after meeting Larry Harvey. He sent her a video of the 1989 Burn. "I didn't know what it was but I knew I had to go." She showed up with a set of silk wings, sixteen feet wide, and climbed to the shoulders of the Man. Wearing them on the night of that Burn, "I felt like I was the protector of the Man and if he was going to be released we had to do it in the best way that we could."[5]

Will, Cait, and I watched from a distance, reluctant to wade into the crowd amid the recurrent dust storms. We were at F Street, atop the four-story scaffolding observation tower that is the signature of our original camp, It's All Made Up. Its gift to the community is this dramatic overview of the playa. Too far from the scrum to share the crowd's expectation or feel the conflagration's heat, from our perspective the Man disappeared silently. Like an imploding star collapsing into a black hole.

I've got to admit, heresy though it sounds, that the burning of the Man is not my main event of the week. For many, it's the kick-off to a marathon party lasting through dawn, but I see the Man's demise more along the lines of the full stop at the end of a sentence. I'm always more sad than elated. Immediately afterward, the massive exit from Black Rock City begins. Pockets of vacant space appear where physical density previously rivaled Bangladesh. Getting through the gate Sunday will take hours; traffic on the two-lane road leading back to Reno moves at a snail's pace.

With the burning of the Man, my spirits change from excited anticipation of the next incidence of playa magic to sober solemnity. It was in such a mood that I headed Sunday night to the marking of the end of the event: the burn of the Temple.

The Temple is to Black Rock City what St. Peter's is to Rome, the spiritual centerpiece. Like most everything at Burning Man, it arose organically. David Best built the first one. He is an artist (he took classes at the San Francisco Art Institute with Jerry Garcia) who started out to be a printmaker and ended up a sculptor. The Temple of Grace in 2014, his sixth, was intended to be his last. But, unhappy with what appeared the next year, he had returned in 2016 to make a final, definitive statement of what he thought a temple should be, in hopes it would be etched permanently into the culture's memory: a medium of healing and forgiveness. I visited his Petaluma, California, workshop a couple of months before Da Vinci's Workshop. He was already at work on what he called, in keeping with its archetypal intent, simply the Temple. I heard his story while he scurried around Petaluma, scouting out construction material.

Best's virgin Burn was 1997. "I was the classic first-time newbie, partied and left, didn't clean up after myself." Over the next two years he built an installation ("nothing of any consequence") and an art car. "In 2000, I had this material from a toy factory. I thought I'd make something out of it. At the time, I was working with a young kid named Michael Hefflin. He was building a three-wheel motorcycle to take to Burning Man but he was killed before, doing 140 miles per hour on a motorbike. He was twenty-seven, twenty-eight. Maybe younger. I didn't know what to do, but his friends all said he'd still want us to go. These kids had never experienced death before. So we built this thing on the playa, and it became obvious that we were building a tribute to Michael. We wrote our goodbyes to him as we were making it. Other people came by, lots of them, and added names of people they'd lost. Then we put some diesel on it and burned it with no ceremony."

David Best sounds offhand about what happened, but his friend Larry Harvey immediately appreciated its significance. The way people spontaneously responded, processing their grief, made him realize that Black Rock City had been missing something. It needed a place to be still, to be reverent, to remember. A quiet space for meditation. Harvey asked Best to create such a place for the next year. "I wondered, what would I dedicate a temple to? Not having any religion, and not being very fond of religion, I thought how in some faiths you can't be buried in a cemetery if you've committed suicide. So since Burning Man welcomes so many things, the most sacred place at the center of the Temple should be in honor of those who've lost someone to suicide." He called it the Temple of Tears. By week's end, hundreds of names of suicide victims were written on the altar. Elsewhere in the structure, thousands had inscribed memorials to people they'd lost.[6]

Temples are now fixtures at Burning Man, with benedictions inscribed on every inch of available space. Forgiveness and pardons, declarations of long-repressed anger, loves remembered, regrets admitted. People bring objects that they're ready to purge from their lives. Like the checkered Arab warrior scarf my friend Matt was wearing when we met in 2013, campmates at IAMU. He's a big, hulking guy, the personification of the Marine he was in Iraq who took the keffiyeh off his first kill. In 2013, Matt was a troubled man. Later that week, the scarf was missing. I asked where it had gone—was he ready for a different fashion statement? Instead, he told me a tale that stirred me to the core of my being. He had deposited the scarf in the Temple, along with a folded American flag, and written an inscription above them: "Here's to the brother I couldn't bring home, and to the brother I kept from going home."

Best was once asked to explain the difference between a memorial, a permanent place where things stay, and his temples. "When I ask someone to forgive a brutal rape," he answered, "and that person puts the rapist's name in the Temple, I can promise them that it won't come back unless they bring it back. I can promise them protection. Your way of ensuring that the pain and violation go away is to burn it."[7]

The stillness of early morning is my favorite time at the Burn. I bike alone, circling the edge of the city. Past the big theme camps, out to where people pitch a tent by themselves or live in a rented van or maybe with a few friends. Folks waking up wave hello as I roll by. Occasionally I'm offered a cup of coffee. But on one such ride-about this year, I insisted my son come along. Will resisted, full of teenage contrariness. But I was resolute, I wanted him to experience with me (or, perhaps, I wanted to experience with him) the Temple.

Best's creation this year was extraordinarily beautiful, inspired by a pagoda shrine he had helped repair in Nepal after the great earthquake of 2015 (an earthquake powerful enough to cause an avalanche on Mount Everest). When the Temple was assembled on playa, the original crane ordered to hoist the final spire 100 feet high wasn't up to the task, and an even bigger one had to be desperately found at the last minute. If you had trekked up the Himalayas to see this Temple, you wouldn't have been disappointed.

Best's style had always been elaborate filigree, computer-assisted cuts produced by what's called a CNC router. No automation and precision this year, however. He opted entirely for hand-built carpentry to underscore the human element. "Woodworking on the scale of Noah's Ark" somebody called it. Scores of Japanese lanterns, each requiring the joining of over five hundred sticks of different sizes, dangled from the beams. The floor plan was unusually open and welcoming. Surrounding the interior altar was a sprawling walled courtyard meant for quiet repose.

The Temple of Promise in 2015, the one that prompted Best to return and make this one last stand, had been heralded as a radical break from romantic structures of the past. It was an abstract, wispy thing—a cross between a huge nautilus shell and a gigantic ear trumpet (the better to hear faint whispers of the soul). Sleek. But problematic. People entering at the front end got squeezed as the channel narrowed. Crowds pushed from behind. Some people loved the Temple of Promise for its striking design, but others experienced claustrophobia and fled. David Best tried to maintain diplomatic neutrality when I asked him about it, but he couldn't hide his displeasure. To his thinking, the Temple of Promise ignored the essential reason for a temple, which is to provide a spacious sanctuary for contemplation.

It failed what I've come to call the David Best test. He worried that prioritizing form over function, design over utility, would become the new norm. So he set out, in 2016, to package his ideal into a final coda. As Best explained to me in Petaluma, "the Temple is about humility." That was the message of his valedictory. He had returned to create an iconic archetype intended as a lasting example for future temple-makers.

When Will and I arrived, the Temple was nearly deserted. Within a few hours throngs of people would be there, clustered in nooks and crannies, staring silently. Two lotus-positioned meditators at the edge of the entrance sat so transfixed that they might well have been there all night. I stopped before a tacked-up photo of an elderly woman, smiling and sweet. "Your last words to me were not to cry on the way home but to sing," was written below. I imagined the woman in the picture on her hospice deathbed, she and the writer knowing they would never again see each other.

Memories of my own mother, who had passed away five years earlier, flooded me as I stood before the inscription. Emotions of loss and bone-deep regret. I had brought a Sharpie pen with me, knowing this year for the first time I wanted to leave my trace behind but with no idea what I was going to say. Now I did. On a beam I wrote, "Mom, I loved you so much more than I ever told you." And I cried, which Will had never before seen.

The words I spoke to my mother went up in smoke, with all the others, as the Temple burned magnificently Sunday evening. David Best et al. had clearly directed their mastery not only to how the structure went up but also how it would come down. A succession of elegant silhouettes framed by flames marked each phase of its collapse. Thick flumes of ash sucked skyward, the residue of sentiment that had been deposited. A genealogy of pain was being carried off by swirling dust devils, perhaps finally loosening the death-grip of sorrow.

The Temple burn is how Burning Man ends. The final curtain call, a finish to a week replete with promises of fresh starts.

Cait departed the next morning, taking Will home to start eighth grade. I remained two more days to help dismantle our camp. The barren splendor of the playa grows more haunting as it depopulates, the night sky more saturated with stars in the absence of Black Rock City light pollution. By Tuesday, though, the allure of the dusty setting had played out. Only one thing was on my mind: a steaming hot shower. My bags were packed, mentally as well as literally. My last lingering campmates circled to say goodbye and, as I turned to leave, they shouted, "We love you, Crow!" For the second time, I cried.

A friend was driving me back to Reno. John Hartman is a third-generation Nevadan, born and bred, who talks volumes about fishing for giant trout on the Piute reservation and prospecting in the hills for gold. In his earlier iteration as a Washoe County police officer, he was assigned to patrol Burning Man. From that vantage point, he witnessed a passing parade of freaks and druggies and sleazy ne'er-do-wells. Then John's life changed. A drunk driver hit him head-on as he rode his motorcycle. He underwent more than thirty operations. His career in law enforcement ended. A few years afterward, a friend gave him a free ticket to come to Burning Man as a civilian. "It's totally different from the other side." Now he's a believer.

We were talking about Fly Ranch as we rolled through the gate and John, who knew it from his law enforcement days, asked if I wanted a peek. He turned north on Highway 34, the opposite direction from Reno, and fifteen minutes later pulled over to the side of the road. A guard, posted at the entrance, roused as we approached. He looked none too happy for being disturbed. Locals don't much cotton to outsiders. I tried to talk us in, describing the book I was writing. He wasn't persuaded. We retreated.

A few hundred yards farther down the road, John stopped

and we dismounted to survey Fly Ranch through a galvanized wire fence. He warned me to pay attention where I stepped; rattlesnakes are everywhere. Sagebrush was in flower, patches of green nestled in the hollows of the hills. While it's not exactly pastoral, you can understand why ranchers ran cattle here and farmers grew alfalfa. It was another planet compared to the playa. "You move from the arid, spare, depleting lunar landscape of the high desert, to the robust teeming-with-life, nourishing wetlands that feel full of possibility," wrote Chip Conley, a board member instrumental in buying Fly. "Yin to Yang."[8]

My return to civilization was more abrupt than advisable. "Decompression" is a skill veteran Burners practice: resurface to reality slowly enough so as not to implode. I floated straight to the top. In a matter of hours, I was back to driving carpool and walking the dog. For a while, I clung to a last vestige of my Burning Man persona, a vintage cowboy hat I bought for $15 in Gerlach. The look, though, didn't play as well in Washington as in Black Rock City. I felt like a field hand walking down K Street. The hat got stashed in a shipping bin, awaiting next year's journey.

5

"On Our Best Days We're Finders"

Cities have the capability of providing something for everybody, only because, and only when, they are created by everybody.

JANE JACOBS,
The Death and Life of Great American Cities

Torrential rains fall as we drive south along California's Highway 1, past Monterey and Carmel. The last remnant of a Pacific typhoon is winding down. Thick fog obscures Big Sur. It's several months since the wrap-up of Da Vinci's Workshop, and I'm bound for the fabled Esalen Institute, where the Burning Man intelligentsia is assembling to brainstorm their future. I've been invited to attend as an observer.

When I mentioned my destination to friends, I was surprised how few knew of it. In the glory days of the counterculture, back when all the stuff was weird and controversial, Esalen was the veritable tent post in the Human Potential movement. A center for gestalt therapists, encounter groups, aspirant Buddhist monks, evangelical vegetarians, holistic healers, massage therapists, drug gurus, body worshippers—the whole New Age mash-up.

A run-down hotel stood on the site when Michael Murphy persuaded his grandmother to lease him the family property (her husband, a doctor in nearby Salinas who delivered John Steinbeck, had originally bought the land to use for a health spa). It was here Murphy founded the Esalen Institute in 1961, naming it after a vanished local tribe who believed everything—the moon, the trees, the breeze coming down the mountain, the pebbles at your feet—possessed emotion and intelligence. The public might know Murphy best as the author of *Golf in the Kingdom* but his celebrity comes from Esalen. A who's who of meta-thinkers eager to rewire mankind's brain started showing up: Aldous Huxley, Alan Watts, Gregory Bateson. Abraham Maslow, whose concept of self-actualization remains the bedrock of humanist psychology, was a regular.

Esalen's earlier fame, as I realized from the blank stares of nonrecognition back in D.C., has been largely eclipsed. The provocative ideas that took seed there have, in the intervening years, gone mainstream. Success has stolen, if not its historical importance, at least much of its first-mover notoriety. These days, Esalen pays its way as a convention center, albeit one with an introspective bent, offering attendees the opportunity "to rediscover the miracle of self-consciousness."

The site is spectacular. Rustic buildings nestle on a sheltered meadow overlooking a coastline *The New York Times* called "the greatest meeting of land and water in the world." Gardens, vegetable and floral, are impeccably tended. Monarch butterflies flutter about. The fabled hot spring baths, perched on a cliff edge, are perfectly positioned for savoring sunsets. And Mike Murphy still walks around with the same twinkle in his eye that prompted a friend, in the early days of Esalen, to observe, "He wants to turn on the world." An intention he and the Burning Man folks recognized in each other.

Murphy came to Burning Man the first time to celebrate his eighty-fifth birthday. He was hosted by the Founders. His son Mac tells the story of their arrival by private plane. Black Rock City does, indeed, maintain an official FAA registered airport during the week of the Burn. Debarking passengers pass through a portal whimsically marked Customs. It's a joke. But when Murphy was asked what he had to declare, he mistook the trolling Burner greeter for a bona fide narc and reached into his shirt pocket to hand over two joints.

I sat across from Murphy, purely by chance and not recognizing him, at dinner my first night. An animated conversation was underway about the compatibility of scientific inquiry and mystic experience. Murphy was jazzed up about biofields, the magnetic fields surrounding cells of the human body. More specifically, whether "high spirits," holy men and women like the Dalai Lama or Mother Teresa, might be vibrating at frequencies different from the rest of us. As a man of science (Murphy had originally gone off to Stanford to become a physician), he believed that there was an answer to be had. To that end, he explained, a crew of Esalen-sponsored researchers had been granted access to the Vatican archives of sainthood testimony, centuries' worth, looking for behavioral markers that might evidence biofields. Such was table talk at Esalen.

Michael Murphy did not reappear after that evening, but his son, Mac, known on playa as Spawn (or Sexy Spawn to women), was very much present. In his welcoming remarks after dinner, he fused the mystique of the place with the timbre of the crowd. "Burning Man and Esalen," proclaimed Murphy *fils*. "Two pillars of virtue in the healing of the world that we're all part of."

The folks here constituted what one could call Burning Man's "establishment," leaders of the Project and pillars of the community. Some non-initiates were mixed in, invited because of special expertise in subjects relevant to the ensuing conversations about, say, mixed-use urban development tailored to artists, or 3-D printing in austere environments, or proprietary digital currency. Whatever the title of the particular breakout session, the essential subject was always the same: How do we seed, sustain, and expand creative community in the real world based on Burning Man principles?

There was a palpable undercurrent of excitement in the hall, a sense that a threshold was being crossed. In keeping with the commitment to spend the next several years listening to the community at large before undertaking any action, Fly Ranch was not explicitly on the agenda. But it was the ten-thousand-pound gorilla in the room. Burners call themselves a people without a state, but now, with the acquisition of the property, they had a homeland.

As long as Black Rock City was temporary, mistakes could be erased. Once assets were sunk in the ground, actions would become more lasting and harder to undo. Fly demanded that the Project raise the level of its game, something the worldly sorts in the room understood. More than a few had started businesses, worked for big companies, consulted for governments. They got it. But the community's institutional memory was rooted in a different place, one where the cultural imperative was individual autonomy. Not institutional resilience. Larry Harvey, who had weathered many such storms, respected this fierce strain of independence better than anybody. He made this clear in his introductory remarks following Spawn's welcome.

Onstage, Harvey's quirky charisma is lost. Rather than play to an audience, he tends to disregard it. He speaks in cryptic fragments, as if thinking aloud to himself in shorthand. For somebody who likes the limelight, he doesn't seem particularly comfortable actually being in it. Here, though, he was with his people. They knew how to read between his lines. The Philosophical Center had convened this session and, as the chief philosophical officer, Harvey set the tone. "There are two things we know about our community," he said by way of preface. "It's factious and it has a great distrust of authority."

That's where the Ten Principles enters, the magnetic lodestone that holds the fractious community together.[9]

The Ten Principles are the closest thing Burning Man has to a doctrinal catechism. A User's Manual, a profession of the faith. Find yourself in a controversy, say whether turnkey camps should be allowed, and the argument will inevitably come down to Talmudic disputations about the *whatness* of Burning Man debated according to the principles.

There's no little irony in this. Their purpose was never to be consulted in close litigious reading. Harvey never regarded them as revealed text. Rather than resolve disputes, the principles were meant to encourage them. He codified the underlying logic to Burning Man without insisting it be logical.

They serve more as provisional glue than a creed. With the sheer unknowability of what might show up at Burning Man being its whole raison d'être, there needed to be some kind of safety valve. You wanted a controlled burn, not a thermal meltdown. If you were going to bet the farm every year, it made sense to hedge the risk with a little insurance.

That was pretty much the point of setting down the principles in the first place, to provide enough adhesive to keep the whole scene from collapsing under its own weight. As the allure of Burning Man caught on, so did the danger that its cultural values would be overwhelmed by crowds of revelers. The code of Black Rock City had always been spread by word-of-mouth, by example rather than dictate. Quantum leaps in the population, however, jeopardized that mode of informal transmission. The Founders started worrying about assimilating neophytes into the culture, so "they wouldn't just come looking for a party but realize what our ethos was." Which was no mean feat, instilling an ethos of responsibility, considering they were trying to rein in a population that was rebellious by nature.

At the same time, the Regionals were starting to take off. Around the country, around the world, inspired Burners were making their own shows, each an independent act of self-creation. Headquarters thought this was terrific. Nobody wanted a Mother Church exercising control from San Francisco. As the fever spread, though, regional leaders looked to the Founders for guidance. Some kind of codification. What, they kept asking, is the recipe for the magic sauce?

Truth was, there were few secrets to spill. Yet there had to be *something* to pass along, to explain how Burning Man processed all that wild-and-woolly mayhem into a civic society. How was the dissonance between maximum personal freedom and communal responsibility reconciled in Black Rock City? In a word, what was this ethos the Founders were talking about, anyway?

Harvey, though, aggressively resisted codification. Steven Raspa remembers the heavy lifting it took to change his mind. Slight of stature and eccentric in style (his beard once wrapped twice around his neck and hung to his waist), Raspa is regarded by many as a saint. These days he's an event producer for the Project, a role that combines his background as both a poet/artist and business consultant. Back in 2004, he was a founding member of the Regional Committee, the council of regional leaders. "Different groups had different intentions but they all needed a general statement of some kind to let people in their areas know what Burning Man was." The only person to do this was obviously Harvey, but at his cantankerous best he refused to cooperate.

Even a minimal agenda once written down, Harvey argued, would corrupt the delicate balance of the culture, which, he often repeated, was ritual without dogma. "He kept insisting that it was an experiment that wasn't done, that it had to keep going forward," recalls Raspa. Still, the Regionals needed help. They were crying out for a foundation to build upon. Raspa et al. persisted and, eventually, Harvey succumbed to group pressure. "I didn't volunteer," he would say, "I was *voluntold*."

Such an unwelcome assignment justified a hospitable work

setting where he wouldn't be distracted. He traveled to Mexico, to the resort town of Mazatlán where he set up shop at a café overlooking the animated central square. "It was a pretty place to write," he recalled. Hunched over a laptop, drinking coffee and smoking cigarettes, he stared out at the parade of passersby. "I must have looked like an albino termite on speed." In the San Francisco offices today, there is a comfortable lounging area named in homage to his public plaza "office." It's called the Zócalo.

He sketched out notes in a schoolboy's notebook, beginning with the premise that informs the whole venture. It's tough to read, written in his cramped cursive scrawl, but twenty years later the words continue to ring true. "My principle [*sic*] intention as an artist is to create a ritual culture which is accessible to people living in non-traditional society. To accomplish this, I have devised, directed and supported art with the following characteristics."

A brief aside. Harvey is a classic pragmatist, in the tradition of William James. James, for those who might have forgotten, was among the nineteenth-century originators of what is regarded as America's foremost (cynics would say "only") philosophy, pragmatism. Where other philosophers, say idealists or dialectical materialists, posit fixed points of certainty in their intellectual universe, pragmatists see continual flux. "Truth" is malleable, not absolute. The proof, as they say, is in the pudding! Harvey often quotes William James to that effect, "Belief is thought at rest."

These pragmatists were taking on eons of religious orthodoxy, celebrating human agency as the end in itself, not the means to something else. There was no ultimate meaning, no divinely scripted destinations. No heaven or hell. Instead of consciousness being innate, pragmatists insist it takes form through ongoing trial and error. From action comes truth, not the other way around.

Experience thus shaped Harvey's principles. He sought to denote behaviors he saw happening at Burning Man that made it successful. Descriptors. As Harvey endlessly repeats, sounding like a weary schoolmaster rendered cranky from explaining the same lesson to dull-witted students, the text isn't written in the imperative voice. Not a single "thou shalt." These aren't commandments or decrees or even best practices. "They do not give permission," he points out. "Nor do they withhold it." Instead, they stake out the parameters of a culture that, should the Project ever devise a coat of arms, would have as its motto "Action Precedes Thought."

Which set up the principles, "the characteristics" cited in Harvey's notebook, for contradiction. Experience, being a hydra-headed monster, meant that the various axioms were not always consistent. Just because something was true in one context, just because it worked at a certain time and place, didn't mean it necessarily *always* worked equally well. When Harvey finally arose from his labors, the pieces fit imperfectly but each rang true.

Like gifting. "It came from the fact that we didn't have any money," he tells me later, back when Burning Man was like a family picnic. He looks at me quizzically, as he so often did, to make sure I understood. "You don't sell things at family picnics." Early on, vans appeared with drinks to buy but since everybody had booze of their own, that was a nonstarter. Later, another vendor showed up with fireworks; when it was made plain to him that this was uncool, he ended up giving them away. Harvey's tactic is to accentuate the positive rather than invoke police power. "Instead of saying 'no commerce,' we said 'this is a gifting culture.'" The notion of what constitutes a gift itself changed over time—behavior counts as well as things—but generosity of spirit remains the constant.

But even something as seemingly value-neutral as gifting can bump into a competing principle. How does it square with, say, radical self-reliance? "Burning Man encourages individuals to discover, exercise, and rely on his or her inner resources." And didn't that, in turn, slightly undermine number six, communal effort? "Our community values creative cooperation and collaboration." Or civic responsibility?

The answer is yes and Harvey's response was "so what?" The role of the principles isn't to inhibit or constrain activity. He wasn't assembling a jigsaw puzzle where all the pieces neatly interlock. Rather, they serve as reliable points of reference, visible fence posts. Activity within those contours is ever-changing in what scientists call a co-adaptive ecosystem. Burning Man is a culture of continuous experimentation but it's not free-range. The principles stake out the perimeters. What's imperative isn't their internal consistency but the dynamics of the whole. The pragmatists would have perfectly understood.

Although in his heart of hearts Harvey considers himself an artist, he'll say in public that he's a social engineer. He regards his surroundings the way a jeweler appraises a precious stone, turning it over and over in his mind's eye to examine each different facet. In launching the Project's Philosophical Center, Harvey summed up the function of the principles by invoking William James's emphasis on consequences. "The relevance of any idea should be gauged by answering a simple question," Harvey wrote, paraphrasing his intellectual mentor. One judged merit not in terms of right or wrong, good or evil. Instead, the critical question to ask is "What real difference will it make if one believes it?"

Harvey returned from Mexico to share his handiwork. Radical inclusion, gifting, de-commodification ("social environment unmediated by commercial sponsorships, transactions, or advertising"), radical self-reliance, radical self-expression ("unique gifts of the individual"), communal effort, civic responsibility ("assume responsibility for public welfare"), leaving no trace ("no physical trace of our activities wherever we gather"), and participation ("we achieve being through doing").

The crew at headquarters saluted his accomplishment. They cheered him for getting it right, not only catching the essence of the culture but, even better, doing it without legislation. But what the hey, only nine principles? That was the joke around the office. Moses descended from Mount Sinai with ten! So, acknowledging his task uncompleted, Harvey showed up the next morning with number ten. "I forgot the most important one," he confessed. He had clearly dug deep, penetrating his sometimes crusty exterior to unearth something touchy-feely. It was as if all his sublimated mushy sentiments burst forth to be bundled into principle number ten, immediacy.

"Immediate experience is, in many ways, the most important touchstone of value in our culture," he wrote. "We seek to overcome barriers that stand between us and a recognition of our inner selves, the reality of those around us, participation in society, and contact with a natural world exceeding human powers. No idea can substitute for this experience."

And, lo and behold, publication of the Ten Principles lit up the switchboard. The Regionals had their take-off point, but what surprised the heck out of Harvey was the thunderous response from the community at large. "When they were published," he recalled long afterward to me, still sounding slightly stunned, "a very remarkable thing occurred." Within a community resistant to any form of central authority, the principles were met with broad acceptance. All those Burners out there, craving a way to replicate the ineffable on-playa experience, that thing "you just can't explain in words," now had a starting point. There was a preliminary vocabulary of what linguistic professors call cultural inter-subjectives, collectively invoked concepts to undergird discourse. Shared buzzwords.

Meanwhile back at Esalen . . . Harvey is addressing the room as chief philosophical officer, encouraging the audience to use the principles for guidance in their subsequent explorations. "Look to the unknown." That was his message to the room. "All will be lost if we pass from a thing that is discovered to a thing that is merely received"—particularly with respect to, and here he used the word "utopia." Which was a not-so-disguised allusion both to Fly Ranch and beyond, infusing Burning Man culture into the world.

The promise of Burning Man has always been ideas. Back in the day, the locus of intellectual exchange was a run-down Edwardian on Golden Gate Avenue where latte carpenters and performance artists talked late into the night. Cultural gypsies. Folks who can no longer afford to live in San Francisco or even the East Bay. The romance and legacy of those conversations looms large. Harvey correspondingly called upon his audience to restoke those kitchen-table dialogues while at Esalen.

Within the structure of the Project, Harvey still remains the looming eminence. Close colleagues, even Marian, often refer to him in mock-formality as *Mister* Harvey. He invoked that mystique in closing. "I like to say that on our good days we're Founders. On our worst days we're flounders. And on our best days"—he milked the moment with a dramatic pause—"we are Finders."

Thus dispatched, the assembled went forth to find.

Over the next days, they shared visualizations of possibility. Much of this came down to place-making, the subtext to a whole slew of topics. How do you create physical spaces that bring out the best in people? Who is necessary to have a creative community? What kinds of structures and design? How do you secure financing? What's the role of technology? As Finders, they were looking for ways to incubate Burning Man–esque spaces. Like good pragmatists, they were open to the wildest hypotheses. The folks summoned to Esalen represented the expeditionary force in the next phase of the campaign to change the world.

Ideas got tossed back and forth with abandon. Did the community need more rules? Can artists survive much longer as artists? Should the Regionals be formal affiliates? How might the authority of the bureaucracy be strengthened without strangling autonomy? Were there solutions to Burner burnout? Will courting philanthropists compromise the culture's integrity? Should Burning Man have an official constitution? Can the Burner approach revitalize poor neighborhoods? Might, heaven forbid, the talk of "changing the world" itself be misguided?

All kinds of potential scenarios surfaced. Micro-lending arrangements between Regionals. B-and-B residencies where patrons hung out with the artists they were supporting. Property models that enabled the community to accumulate equity in real estate. Calls to action to mobilize civic welfare endeavors. Peer-to-peer education banks.

A topic that took root was digital currency. One of the academic wizards in the field, John Clippinger of the MIT Media Lab and founder of the Center for Internet and Society at Harvard Law School, was at Esalen in his role as co-author of a book titled *From Bitcoin to Burning Man: The Quest for Autonomy and Identity in Digital Society*. Details about blockchains, verification algorithms, and distribution firewalls were lost on most attendees. What the crowd *did* get was the macro-concept: the theoretical possibility for the Burning Man community to create its own cryptocurrency. Or, on a smaller, more manageable scale, a beta test with one of the big factory-space art collectives, like the Generator in Reno. The propositions were intoxicatingly seductive.

During a daylong breakout session, in the serenity of a secluded garden bejeweled with a statue of Buddha, I sat in with a group that included Harvey. Our nominal facilitator was Jennifer Raiser, San Francisco presence extraordinaire and also a Project board member (and treasurer), as well as the author of a stunning coffee-table book, *Burning Man: Art of Fire*. A woman of polished grace, she provided bottles of good wine and gourmet snacks for our afternoon break. "Ahh," Harvey said to me in an aside, "with Jennifer around, you always live well!"

Talk got around to Fly Ranch, about how it fit into the grand scheme. "Never underestimate the power of place," Harvey intoned, an observation worthy of a former landscaper. The setting, he suggested, might itself be turned into a narrative, with elements arranged so a story unfolded to visitors as they passed through. "Restore nature as the catalyst of mystery and wonder" was his recommendation, which prompted an extended riff about the measure of art, of myth, of personal relationships as transports into the unknown.

One preeminent place-making function of Fly, Harvey envisioned, was to draw people unlikely to rough it in the desert. To call them movers and shakers, thought leaders, celebrities with name recognition would be too crass and obviously self-serving. Still, I heard the voice of the promoter in this approach, eager to expand his base. Expose folks with clout to the transformative power of the Burning Man virus, then let them scatter the vectors of contagion.

One such person of influence, already contaminated and fully enamored of the ethos, sat among us in the Buddha Garden group. Ping Fu, *Inc.* magazine's Entrepreneur of the Year in 2005, carries herself in a reserved, composed manner that gives no hint of what she's accomplished. Her extraordinary life has led from China (during Mao's Cultural Revolution, as a young girl, she was separated for years from her parents) to the United States (which she entered penniless and unable to speak English) at age twenty-six to a master's degree in computer science from the University of Illinois. (Among her contemporaries was Marc Andreessen, the founder of Netscape.) Ping subsequently founded her own company, Geomagic, a pioneer in 3-D printing that launched a new industry of "personal factories" that changed how molds and tools are manufactured. Customers included Ford, Boeing, Timberland, and even the National Aeronautics and Space Agency. "I was handed few advantages in life," she writes in her compelling memoir, *Bend,*

Not Break. "I possess no extraordinary talents. I simply was born with the curiosity to learn, the tenacity to make a better life, the desire to help others, and a great deal of resilience."

Her virgin Burn was 2010. Chip Conley, the Project board member who spearheaded the fund drive for Fly, crossed her path in entrepreneurial circles (he founded a chain of boutique hotels). His birthday party was going to be celebrated at Burning Man that year, and he invited her to attend. "I rushed around frantically to see the art, knowing it was going to disappear." Subsequently, she became more relaxed. "I went from being a visitor marveling at how wonderful everything was to a citizen. I started thinking of Black Rock City as *my* city." She was struck with how, when she'd meet someone off playa who identified as a Burner, she felt instant kinship. "I trust that our values are aligned."

At the end of that first Burn, Ping was brought to Fly Ranch. "A blissful feeling" was her first impression. Several years later, around the time she sold her company, the owner of Fly Ranch was hinting he might sell. The price had risen dramatically since that 1997 Burn, when the dream of owning the land was first seeded, but so had the leadership's confidence that money could be found. A discreet, below-the-radar development campaign began, and Ping was identified as one of several dozen potential donors. They were not approached by professional fundraisers; that option had been explored and then quickly vetoed by the Founders. They were going to have none of the quid pro quo conventionally used to entice big contributions. No naming rights, marketing connections, or special privileges. Instead, appeals were made on behalf of the community for the sake of the culture.

As talks ensued, Ping had her own ideas of how Fly might be used. Burner-inspired architecture, for example, or inventing new strains of agriculture bred to thrive in the desert. She was tempted to make a substantial gift. But not yet persuaded. Then at a gathering, an exercise was conducted: If "play" were the word for Burning Man, what would a comparable word be for Fly Ranch? Ping thought of the soothing comfort of the muddy hot spring, an emblem of the assurances of acceptance the culture offered. The word came to her. "Love." Ping saw Fly as a place to nurture what she calls Big Love, the love that "makes you give to things that are bigger than your life."

There was a second reason she decided to give. As a tech entrepreneur on the factory floor or as an economic adviser to the prime minister of the United Arab Emirates, Ping understood the importance of there being a woman in the room. "A woman brings the gentleness and softness of female energy into an organization. I'm not a feminist but I feel like the world needs a balance." Other female donors had been sought for Fly Ranch to no avail. She considered it imperative there be one. "The germination of love, its longevity, requires males and females."

Returning to Washington, I wondered how successful the Esalen meetings would ultimately prove. Part of me doubted that there had been much lasting accomplishment. Rich dialogues, to be sure. Considerable networking. Healthy venting. An extended moment to breathe and contemplate. All of that transpired. But in terms of tangible check-off items on an action agenda, I hadn't seen much. When I doubled back with Project leaders, they largely agreed. True, there was no new playbook. Just convening such a session, however, was an experiment that had gone well. Who knew where it would lead?

What *did* definitely emerge from Esalen was a broad consensus eager to deploy the community's enormous reservoirs of talent and energy to take on new initiatives. The organization's capacity was growing. Acquiring Fly demonstrated the presence of philanthropists who believed in the cause enough to put up their money. The rank and file were primed. That was half the battle. The harder part remained: committing to a course of action. Deciding to what ends in the "default world" those resources should be devoted. "We need," as one person characterized Burning Man's evangelical manifesto, which awaited writing someday, "to be able to talk Wall Street, Main Street, and the Esplanade."

6

Chaos and Control

If you're not doing some things that are crazy, then
you're doing the wrong things.

LARRY PAGE, Google cofounder

To fuse Wall Street and Main Street with the Esplanade, Burning Man must graft aspects of its culture onto real life. No element is more important to that culture than gifting. This is the
linchpin that prevents the assorted cogs and wheels from going
off the tracks. One key to expanding the relevance of Burning
Man beyond the playa, of changing the world, is figuring out
how to integrate its practice of gifting into conventional settings.

In a recent book, *Thank You for Being Late, New York Times* columnist Tom Friedman takes on the prevailing malaise of Western civilization, the feeling so many people have that so much of their lives is spinning out of control. Rather than despair, he sees reason for optimism in the analysis of social theorist Dov Seidman, who predicts the emergence of a new paradigm. If the industrial economy was about hired hands and the knowledge economy about hired heads, then the technology revolution heralds a "human economy" that will create value with hired hearts, what Friedman calls "all the attributes that can't be programmed into software like passion, character, and collaborative spirit."

Gifting, when you get down to it, is the heart of the matter.

If, as Harvey declared when he came back the morning after to add one more principle to his list of nine, immediacy is the energy that keeps all the different balls in the air, then acts of giving constitute the intelligence that arranges the pieces. Gifts, grand and small, are the playa's counterpart to physics' pi mesons, its quarks of information that hold everything together.

Were there a single aspect that most makes Burning Man historically noteworthy, one über-departure from the norm that future chroniclers will cite as the inflection point of a new genus, my hunch is that it will be gifting. To be sure, lots of playa phenomena are dazzling. Take your pick! But what carves out the great divide that marks it as a unique species is the intertwining of gifting into Burning Man's collective DNA.

Back in the analog day, when newsletters were the medium of communication, Larry Harvey wrote in the literary persona of Darryl Van Rhey (an anagram of his name with a *d* and *n* taken from his middle name, Dean). Why this mystery man nom de plume? Harvey was eager to tell me. "As Burning Man's resident intellectual, I really didn't have much of anyone to talk to." In need of a worthy interlocutor with whom to dialogue, he invented a furtive eccentric, positioned as the community sage. "I informally floated a biography, claiming that Van Rhey had been born on a train in transit through the Benelux countries and was a citizen of all three, since no one could identify which of these three nations he was actually born in. A cosmopolitan from birth." Insiders knew Van Rhey's true identity, to others he was real.

One Van Rhey disquisition, formatted as a Socratic dialogue, deconstructed gift giving. Markets are impersonal, Van Rhey argued. "Let's say I sell you my hat, when you buy it and I take your money, our business is over." Now, that's a good way to conduct commerce on a macro-level or to aggregate wealth. It's what capitalism is all about. But the same can't be said with respect to human relations. Markets aren't good conductors of intangible assets like trust and reciprocity. Moreover, Van Rhey polemically pointed out, a market economy by definition isn't sustainable over the long term; it must inevitably deplete resources. Ultimately it consumes its seed corn. A gift economy, on the other hand, passes value "from heart to heart," compounding social capital. "Gifts are," explained Van Rhey, "quite literally, bearers of being."

"Regard *yourself* as a gift," he entreated. Be a benefactor. Donate your talents, your empathy, your consideration. Doing so will make you feel real. The gift needn't be big, it can be embodied in modest actions. "We've found that when people join together in this way—not just to share among themselves, but to create a greater gift—it generates a kind of social convection current," concluded Van Rhey. In the same way that a thermal column, say the eye of a hurricane, accumulates power by sucking in whatever it touches, so does gifting compound human energy. Gift one produces gifts two and three, which in turn produce gifts four, five, six, seven, eight, nine . . . ad infinitum.

Van Rhey is more mystic than socialist. He's not talking about communalizing the means of production or the distribution of goods but, instead, establishing a space where transactions—be they barter or specie—are suspended. In the language of Burning Man, a "de-commodification zone."

Van Rhey's theoretical underpinnings came from a book published a few years after Harvey's first Burn, *The Gift: Imagination and the Erotic Life of Property*. "A life-changer," one reviewer called it. Author Lewis Hyde was an obscure poet. By profession, a self-described "scholar without institution." The focus of his attention is the nature of artistic inspiration, "a thing we do not get through our own efforts but is bestowed upon us." A gift from the muse. The result, a painting or a poem, becomes a second-generation gift, rendering beneficence. Worth is measured not by the price fetched in the market but by the value it contributes to people's lives.

"Any exchange, be it of goats or ideas, will tend toward gift if it is intended to recognize, establish, and maintain community." Gifts constitute "anarchist property," expanding in worth as they move from person to person but belonging to nobody. The emotional rewards are akin to an erotic act, the appetite for bestowing them is correspondingly compelling. "Each donation an act of social faith," writes Hyde, "the binder of many wills . . . nourishing those parts of our spirit that are not entirely personal but derive from nature, the group, the race, the gods."[10]

One can imagine Van Rhey's brain exploding with the turn

of each page. Here, expressed in the language of spirit, was the epistemological foundation for all those transformational moments, those life-changing experiences, at Burning Man.

If Burning Man is, indeed, a testing ground for new forms of social organization in the emerging Anthropocene era, acts of voluntary offering and grateful receiving will be at the core of whatever evolves. This faith in gifting, in the abundance of human generosity, is the metaphysical dark matter that supports the sprawling cultural architecture. Herein lies the magical pixie dust those Regionals were seeking. Without gifting as its cornerstone, Burning Man would be just another spectacle. Imagination might still reign and inhibition disappear, but should the cash nexus ever ensnare a critical mass in Black Rock City, the allure of the event as a "world changer" would diminish.

The critical question, then, is: Can gifting be exported without being compromised? In kingdoms of the weird like Black Rock City, liminal zones "betwixt and between the familiar," there's a place for such seeming aberrance. But what happens when you leave the realm of weeklong festivals or socialist summer camps? Can Burning Man's version of gifting get traction off playa?

Silicon Valley is an apt place to look for overlap with the Burning Man ethos. The early digital entrepreneurs, the Google boys et al., recognized how their start-ups mirrored Burning Man. They saw themselves as a community of cohorts fostering creative collaboration. "So much of the culture of really innovative tech companies," observed a student of both high tech and

Black Rock City, "is materially impacted by the inspiration that Burning Man has provided."[11] Facebook's corporate mantra in the beginning, "move fast and break things," would be equally applicable on playa.

Did some random property in the water account for the kinship of Burning Man and the tech community? Or were there systematic similarities each recognized in the other? Professor Katherine Chen, a sociologist at City College of New York, wrote her book *Enabling Creative Chaos: The Organization Behind the Burning Man Event* as an extended inquiry into the organization's approach to governance. Most volunteer-run enterprises—organizational theorists like Professor Chen call them "mission driven"—end up disintegrating. Sustainability is rare. Yet Burning Man, against all odds, thrives. Attendance at the event holds steady because of limits imposed by the Bureau of Land Management, but the community continues to scale worldwide. Moreover, this exponential growth has occurred without distorting the founding values.

Academics who study these things say such success is highly aberrant. Mission-driven entities are great for generating gung-ho enthusiasm but, over time, this typically produces institutional confusion. At the other end of the spectrum, administrative bureaucracy is efficient and rational but at the cost of stifling idealism. Few organizations can fuse together both styles. When volunteer idealists try to implement a structured business-oriented approach or, alternatively, an established bureaucracy seeks to become more participatory and transparent, bad results usually follow. Organizations have to opt for one model or the other and accept the consequences. Burning Man, however, has found a way to walk this tightrope.

Professor Chen explains how it manages to optimize both systems. Its governance model evolved to be able to coordinate multiple moving pieces at ever larger scale while still remaining sufficiently responsive to satisfy the community's commitment to radical autonomy. This wasn't premeditated. Problems showed up, ad hoc responses were reached through arduous application of the Ten Principles. When solutions worked, they became institutionalized as standard operating procedure. Pragmatism was writ large as institutional memory.

Were Burning Man a unique exception, a black swan, students of management wouldn't find it so interesting. Chen's implied thesis, however, is that aspects of the Burning Man approach can be replicated among other operations, particularly operations that start out with more social capital than financial resources. Think of nonprofits like Teach for America or a handful of techies in a garage workshop.[12]

This ability to combine bureaucracy and collectivism is something business-school types haven't before seen. So wrote Rakesh Khurana, professor of leadership and development at the Harvard Business School (and then dean of Harvard College). "From open source software to the election of Barack Obama," he noted in appraising Chen's book, "organizational research has not yet come to terms with the conceptual and theoretical underpinnings of these new organizational forms." Her analysis of Burning Man, he concluded, contains "implications for organizational theory and managerial practice" applicable to the rise of pop-up entities that are part commercial and part communal.[13]

This hybrid form is a harbinger of things to come, an adaptive response to new historical circumstances. Social media campaigns and entrepreneurial start-ups exemplify the first wave of "work" groups held together by emotional allegiance as much as material compensation. The ultimate reward may be an electoral victory or an IPO. But those things are never guaranteed. What keeps overworked volunteers and underpaid employees loyal? Chen concludes that in Burning Man's case, the answer is shared experiential payoffs. Positive emotional moments. A richer sense among the participants of their brotherhood and sisterhood. In short, the gifting of one's self to a grander purpose than personal ambition.

It's worth repeating one more time what we're talking about. Business schools teaching management consistently emphasize the difficulty for a company to evolve from a band of believers into a complex bureaucracy. Burning Man is a conspicuous exception to the rule. As such, it may well be a model with lessons to teach to other hybrids.

Like, say, to Google in its formative days, whose corporate culture is saturated with Burning Man.

In late August 1998, with the company still unincorporated and operating out of three bedrooms and a garage, the original Doodle unexpectedly showed up on Google's rudimentary homepage. Founders Sergey Brin and Larry Page get credit for inserting a Burning Man figure onto the site, an "out of the office" notice informing visitors that nobody was manning the shop. The entire staff had gone to Black Rock City.

Then there's the tale of how Brin and Page hired Eric Schmidt a few years later. After securing a $25 million financing round, the big VCs who lent them the money leaned on the twentysomething boys to hire a CEO for "adult supervision," which they resisted unless Steve Jobs would take the job (he was the sole person they'd accept). The search dragged on for more than a year. "Larry and I managed to alienate fifty of the top executives in Silicon Valley," recalls Brin. They sought qualities rarer and more elusive than technological or managerial expertise. What they wanted was somebody who could sublimate his or her controlling ego into the fabric of the company. "Someone who could discipline Google's flamboyant, self-indulgent culture without wringing out the genius."[14] Finally, in Eric Schmidt, twenty years their senior and with experience running big operations, they sensed they might have their man. But doubts remained. So, to test his mettle under fire, to see just how collaborative and flexible and comfortable with eccentric craziness he *really* was, they took him to Burning Man. He passed.

Then there's how Google Earth was inspired by Black Rock City (as Google Maps had beta-tested there). So this tale goes, Larry Page was on a flight over Nevada when he pointed out to his companion, Google's chief business officer at the time Nikesh Arora, where Burning Man occurs. Then, legend has it, Page zooms off into the zone and passionately starts envisioning a colossal project. "Wouldn't it be cool," one can imagine him saying, "to send planes and helicopters crisscrossing the United States, taking images at low altitude." Arora whips out his calculator for some quick feasibility calculations, and . . . voilà . . . three-dimensional mapping was born. Gifting was central to this enterprise. "The power of Google," explains the chief engineer of that project, "is that they don't do all the work. People posting content do. The same is true here at Burning Man. Citizens create the vast majority of things."[15]

The residual influence of Burning Man on Google was obvious to Stanford professor Fred Turner. Writing in 2006, after

The Founding Founder, Larry Harvey (2003). A mythic trickster who combined the qualities of a visionary, philosopher, politician, and social engineer.
(Photo credit: Michael Mikel)

A rare photo of the six Founders together, shot at the 2013 Burn. LEFT TO RIGHT: Will Roger, Crescent Rose, Michael Mikel, Larry Harvey, Harley K. Dubois, Marian Goodell. "The six of us were married to each other all those years. We learned that we made better decisions as a group than individually." (Photo credit: Courtesy of Karen Kuehn)

John Law (LEFT), a foundational organizer in the earliest years, wanted to shut down Burning Man after the catastrophes of 1996. Larry Harvey didn't. This photo shows the two men in the midst of their intense disagreement on the final night, after which they barely exchanged words. (Photo credit: Maggie Hallahan/MHPV)

The founding sisters of 3SP: Kathy Baird, Cait Clarke, Lesley Stein, and Marie Blakey. Women in leadership are more likely than men to challenge standard approaches, create an atmosphere of continual improvement, get others to go beyond what they originally thought possible, and quickly recognize situations where change is needed.
(Photo credit: Lesley Stein)

"The last mile." Heading toward the Burn along Nevada Highway 447.
(Photo credit: Kenny Reff)

Black Rock City (2014). "An urban design to abet a sense of communal belonging and lead to social interactions." (Photo credit: Kenny Reff)

The art theme in 2016 was Da Vinci's Workshop, with the Man an homage to Leonardo. Harvey wanted the community to think of itself as cultural patrons whose creative efforts and financial support could launch a second Renaissance. (Photo credit: Vanessa Franking)

2017 Man. Against opposition, Harvey put the Man on the ground rather than on a tall base as in the recent past. His intent was for visitors to gaze upward from its feet and feel physical awe. "I'm convinced that people need that experience, desperately. But without recourse to the supernatural." (Photo credit: Kenny Reff)

@EARTH#HOME (detail), a love letter to the planet from artists Laura Kimpton and Jeff Schomberg, the 2016 installment in their Monumental Word Series.

(Photo credit: Vanessa Franking)

Phoenicopterus Rex, the world's largest plastic pink flamingo, created by Josh Zubkoff (2017). "Probably the most absurd lawn ornament ever created." (Photo credit: Vanessa Franking)

Hybycozo, by Yelena Filipchuk and Serge Beaulieu. Art as the intersection of math, science, technology, material, and light (2018). (Photo credit: Vanessa Franking)

The Tree of Ténéré, created by Zachary Smith, Alexander Green, Mark Slee, and Patrick Deegan, offering "shade to wanderers, adventure to climbers, and infinite possibility beneath its 25,000 LED leaves" (2017). (Photo credit: Vanessa Franking)

R-Evolution, the final sculpture in Marco Cochrane's Bliss Project, which asks, "What would the world be like if women were safe?" (2015). (Photo credit: Vanessa Franking)

The family as Burners. The author with wife, Cait, and children Amelia and William (2014). (Photo credit: Kenny Reff)

The Temple (2017). David Best came out of retirement to offer the community a final archetype for future reference of the temple as sacred space. (Photo credit: Vanessa Franking)

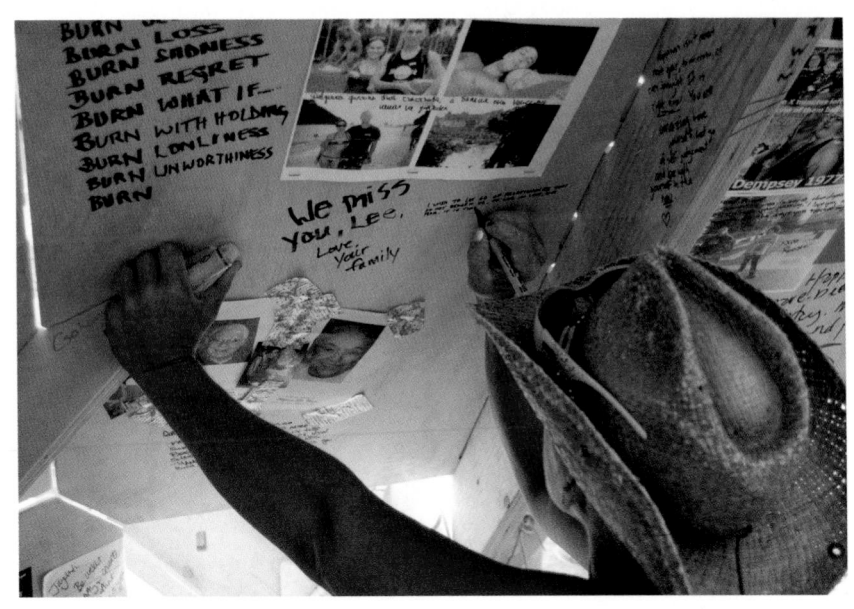

Leaving an inscription on the wall of the Temple. Affirmations of love, grants of forgiveness, pleas for pardon, messages of hope (2015). (Photo credit: Kenny Reff)

the company had settled into the fabled Googleplex, he contrasted the lobby of the main building with what you'd find at, say, AT&T or IBM. "The cool blond wood and carefully recessed lighting which have marked the power of industrial firms for decades have disappeared. In their place, plain white walls are posted with some two dozen unframed photographs of giant sculptures set out in a flat, white desert and of fireworks exploding over the head of a giant neon stick figure."[16] Images from Burning Man.

Anecdotal history aside, implications abound that Burning Man culture deeply filtered into the Google bedrock. Much like the Ten Principles, Google's early ethos was formally inscribed in a manifesto, "Ten Things We Know to Be True." More than your typical aspirational mission statement, these adages offered guidance: Focus on the user; do one thing really, really well; fast is better than slow. They didn't tell their associates *what* to do, "not prescriptive," as Harvey would say, but they did clarify how to evaluate whether or not a venture was sufficiently Googley.

There was also a commitment right from the get-go to think big and follow creative instincts, conventional wisdom be damned. Burning Man acts this out from the artistic side, Google as computer engineers. Each displays limitless confidence in the ability of its community to accomplish seemingly impossible feats like building a temple worthy of Kathmandu, or linking every web page on the Internet into the mother of all search engines. More important than bravado in making these things happen is the organizational approach, hyper-socialized collaboration.

Academics have coined a term for these collective workspaces pioneered at Burning Man and Google: commons-based peer production. Commons-based peer production significantly differs from solitary artists in a studio or a sequestered corporate R&D team. The social component in the process is inseparable from the final result. The key to engaging a bunch of people passionately in a project, as volunteers or at levels way below their pay grade, is the worthiness of the challenge ("change the world") but also the subjective pleasure, internal satisfaction, affirmation of self—just plain ol' fun—experienced as part of a simpatico band of brothers and sisters.

"We had all these amazing people around us," recalls Susan Wojcicki about those late nights with Page and Brin living off pizza when she was employee number 16. "Their goal wasn't to become famous or make money, it was to do something that was meaningful for the world because they had a passion, they found something interesting and they cared about it. I mean, it could be ants or it could be math or it could be earthquakes or classical Latin literature."[17]

Commons-based peer production is all about aggregating cognitive capital and amassing collaborative effort. The role of the leader in this context is not that of final decider but rather curator of the community. As with open-source code, as with Harvey's pragmatism, there are no final solutions in this formula but only provisional answers crafted through the input of multiple voices. The endgame is continuous adaptation.

For Burning Man this means tearing down Black Rock City, carting everything away to "leave no trace behind," and then building a whole new iteration the next year. Google doesn't wipe the canvas clean, but deep in its DNA is the same legacy of pursuing the next big thing even at the expense of disrupting its own protocols. This was made clear in the letter the Google founders wrote on the eve of the company's IPO. In it, to warn investors what they were buying into, Brin and Page declared they would encourage engineers to devote a goodly portion of their hours, 20 Percent Time, to projects of their own choosing, "working on what they think will most benefit Google." Not what their bosses bless.

Embracing the voice of the community, as Page and Brin learned at Burning Man, empowers individual members to address opportunities that might otherwise be ignored or dismissed, even by charismatic leaders.

Take this example, which Larry Page recounted at a gathering of the global elite at the World Economic Forum in Davos. When asked to predict what lay ahead on the technology frontier, he advised the audience to discount his answer. Here's why. "When I was heading up Google X (the company's semi-secret research facility) a few years back, there was one little Artificial Intelligence project there." He held up his fingers to demonstrate just how teensy-weensy it was. Having been educated as a computer scientist in the 1990s when everybody knew AI couldn't work, Page didn't pay it any attention.

Still, Jeff Dean, one of Google's top computer scientists, had carte blanche to keep working on the project code-named Brain. One day, he comes racing to Page, deliriously waving an image that was, plausibly, a cat. The printout had been rendered from the neural network of 16,000 processors Dean had programmed with deep-learning algorithms able to recognize the different aspects of cats from ten million random YouTube videos. "Okay, that's very nice, Jeff, whatever," was Page's unimpressed response. "Now go away and do your thing." Fast-forward a few years. "Now Brain probably touches every one of our main projects

ranging from Search to Photos to Apps to everything we do,"
Page confessed to the Davos audience. "This Revolution in Deep
Nets, neural structured learning, definitely surprised me even
though I was right there when it was happening, so close I could
have thrown a paper clip from my office and hit Jeff."

A company doesn't become a community simply by declar-
ing itself so. Transparency and collaboration are necessary but,
as logicians say, not sufficient. Something more has to be added.
As per Burning Man's example, gifting is the magical catalyst.
Correspondingly, acts of giving have been fundamental to Google from its outset.

The aforementioned Susan Wojcicki, today the CEO of
YouTube (one of the entities composing Alphabet, Inc., Google's multi-billion-dollar holding company), was exposed to the
gifting culture when Brin and Page sought to hire her away from
Intel to become their first marketing manager. Her connection
to the boys was established by then; it was her garage in Palo
Alto they were renting. "I was older than them and I had a business school degree and I had actually worked somewhere. So I
was a little cautious about working for two students. That's how
I saw them at the time: students who were doing their first company." She felt obligated to confide a secret that might dampen
their enthusiasm: she was four months pregnant. No problemo!
Without hesitation, they volunteered on the spot to build her a
nursery. Touched but knowing they were cash-strapped, she declined the offer. A few years later, though, thanks in part to her
ardent advocacy, a company-run childcare facility opened in the
Googleplex (not without controversy: attendance was limited,
waiting lists long, fees subsidized but not cheap).

Now, nobody is going to suggest for a moment that Google

is a decommoditized zone. The benefits employees get—barista-staffed coffee bars, gourmet food, shuttle service, sleep pods, medical checkups, hair salons, massages, workout sessions, 110-day paid maternity leave (60 days paternity)—aren't gifts from the heart. Rather, they're perks designed to keep workers happy and on the job. "The goal is to strip away everything that gets in our employees' way," explained Eric Schmidt. "Let's face it: programmers want to program, they don't want to do their laundry. So we make it easy for them to do both."

Even granting ulterior corporate motives, narrowing the divide between the work and personal sides of an employee's life constituted an act of generosity that Google was among the first to offer. Blurring those boundaries might be an investment in productivity to keep engineers at their drafting screens, but it was also a gift. To be sure, there were self-interested reasons the company wanted to be "the happiest, most productive workplace in the world." The generosity here, three free meals a day, is strategic. A happy environment is the trade-off for commitment and loyalty. At the same time, there's no denying that reciprocal good karma is being traded back and forth. Perhaps not at the same decibel level as at Burning Man but Google's offering credible expressions of generosity that contribute to a high-trust environment.

"I think of how far we've come from where workers had to protect themselves from the company," wrote Laszlo Bock, Google's fabled former head of People Operations. "My job as a leader is to make sure everybody in the company has great opportunities, and that they feel they are having a meaningful impact and are contributing to the good of society." Bock's imperatives square with Burning Man's notion of community.[18]

Larry Page once floated the idea of spinning off a Black Rock City–like zone where entrepreneurs could try out wild ideas. During a three-hour presentation on new products offered at Google's annual developers meeting in 2013, he conceded that technology was not always a blessing. Human nature can't adapt as quickly as technology advances. "There's many, many exciting and important things you could do that you just can't do because they're illegal, or they're not allowed by regulation, and that makes sense, we don't want the world to change too fast." What's the answer? How about a physical venue for tech testing. He cited Burning Man as a model. "We should have some safe places where we can try out new things and figure out the effect on society without having to deploy it to the whole world."[19]

Much has changed at Google in the intervening years. But Brin and Page continue to come to Black Rock City and there's little question that the company in its formative phase was simpatico with Burning Man, as was much of Silicon Valley. "With its emphasis on teamwork, flow, peer production, meritocracy and reputation building, Burning Man culture clearly celebrates the values and practices common to high-tech production."[20]

Gifting, as an expression of mutual respect, is likely to become a signature management tool in the forthcoming economy where capital will be predominately human and collaborative-contributors the unit of labor. Amped-up people-power will constitute the critical differentiator of a successful company. The social networks of commons-based peer production rely on value-based relationships, which suggests how much "hearts" must be engaged. Unlike in capitalism, enterprises will seek to engender feelings of self-realization among a workforce who think of themselves as colleagues and peers.

Burning Man offers an exemplary model of how to operate in the human economy. Few can match its record in prioritizing personal authenticity without succumbing to organizational chaos. Or in affording stakeholders "a special feeling of worth from membership in a prized community." In the next decades, the ability to quickly mobilize temporary pop-up groups will become an increasingly effective tactic for action, whether commercial, philanthropic, or political.

Burning Man has demonstrated singular success in doing this.

Without any central direction, entities within the community have spontaneously organized themselves throughout the history of Burning Man. Burners Without Borders, Black Rock Labs, the Black Rock Arts Foundation—all operate beyond the playa on the basis of volunteerism. The Regionals similarly took form this way. The sum effect of the culture is often described as a "permission engine," inspiring people to seize initiative on their own accord without being burdened by chains of authority.

"It's not your job," goes a playa adage about jumping in when work needs doing, "it's your turn." This state of mind, the linchpin to Burning Man culture, is the ultimate gift. Feeling responsible and empowered to act on behalf of the greater good.

III

BURNING MAN'S
UTOPIAN VISION

7

Art as a Trojan Horse

Burning Man is art that is generated by a way of life and seeks to reclaim the realms of politics, nature, ritual and myth for the sake of art. This is art with a utopian agenda.

LARRY HARVEY

At its outset, Burning Man was *not* about the art. In its present iteration, nothing is more important.

Playa art is the universal touchstone everybody relates to, the feature that legitimates the event to skeptics. Indeed, it is not wholly mad to suggest that as Florence was to the Renaissance, Black Rock City is to a twenty-first-century cultural movement yet to be named, with art at its core. The tail has come to wag the dog.

This emphasis on art, however, arose as a political tactic. An exercise in disinformation. After the first few years in the desert, Larry Harvey started telling anybody who would listen that Burning Man was an arts festival even though he hadn't previously considered it such. This was a ploy, unadulterated spin. His intent was to confuse the authorities, who absolutely *loathed* this annual gathering, which they likened to devil worship. Harvey's strategy was to change the narrative by rebranding the event.

"Sure," one can hear him saying, agreeing with the Bureau of Land Management and local government officials that Black Rock City was a whack show. "Damn right! It would drive anybody crazy. Sometimes, it drives me crazy. But the thing is, these are *artists*! As you and I both know, they're not normal people. They're nuts by definition." That was Harvey's line. He wedged a membrane of redemption into Burning Man that the authorities, and later the media, grudgingly conceded. The Feds backed off a bit. Reporters found a more favorable story line. Tactical brilliance!

The thing was, propaganda to the contrary, very few Burners at the time thought of themselves as artists. There was costuming and prank performance, dressed-up dummies and assorted ironic objets d'art like a car flattened by the 1989 San Francisco earthquake. Stuff that, maybe if you stretched the point, might pass for Dada. Nobody, though, was coming out to the playa to do Art, capital A.

How things changed over the course of several decades!

The Project now budgets more than one million dollars annually to support art projects in various venues and iterations.

Grants are made for work not even destined for Black Rock City. Patrons buy Burner art, municipalities commission it, real estate developers install it as the centerpiece to developments. "Burning Man is used as an adjective among art directors," reports a big-time set designer. "They'll say, 'Can you make it a little more Burning Man-ish.'" No less a custodian of American taste than the Smithsonian has weighed in, devoting the entire Renwick Gallery to "No Spectators: The Art of Burning Man" in 2018.

With the sky literally the limit, artists have seized the challenge to go bigger and bolder. Marco Cochrane, arguably Burning Man's Michelangelo, was blown away the first time he saw the size of the stage he could sculpt upon. Until that moment in 2007, his medium had been three-quarter-size maquettes, working models for somebody else's sculptures. Three years later, he brought to the playa the first installment of what would become his signature series of monumental women in exuberant poses.

Bliss Dance ended up eight times life-size: forty feet tall (a later piece would be fifty-five feet), its seven-thousand-pound body seemingly balancing on one leg (seven I-beams of hidden support were buried below the surface). To create it, he had to learn to weld (wire-mesh skin was stretched over a steel anatomy) and consult structural engineers (the piece needed sufficient tensile strength to withstand gusty desert winds).

Bliss Dance wasn't big just to be big, though. Size was in service to mission. When Cochrane, the son of Berkeley hippies, was seven years old, a neighbor girl was raped outside his house. The effect of this event left him an ardent feminist, the sentiment at the core of his art. His works are expressions of women secure enough to glory in their bodies, women powerful enough to feel free. Like *Bliss Dance*. "She's confident to exude that edgy feeling of being safe enough to be naked and dance with full gusto," he explained when we spoke. "To me, that's the energy that's missing in the world. Women spend pretty much every day wondering if they're in a situation where they're going to be raped. They have to think about it all the time." His work is intended to redress such vulnerability by reversing the "victim" perspective. And who better to inspire women (with support from sympathetic men) to be heroic than a nude giantess?

There are some within the Burning Man arts crowd, to be sure, whose praise is respectful but reserved toward Cochrane's work. They see it coming from a "very old, over-educated" European tradition rather than being quintessentially Burning Man. "I'm in awe of his ambition and follow-through," is how one respected artist puts it, "but I wish he had a fat girl in there instead of an idealized figure."

Among the playa populace, though, there is no such ambivalence. *Bliss Dance* and his subsequent works (*Truth Is Beauty, R-Evolution*) are showstoppers. Indeed, to fans and funders, the raw charisma of these figures represents Burning Man art at its most powerful. Back in storage after the event, *Bliss Dance* had a growing reputation that called out for public display. She's ended up in Las Vegas since 2016, purchased by MGM Grand to preside over a pedestrian park through which some twenty thousand visitors daily pass in their stroll from casino to casino. If there's irony in this, a feminist statement in a scene hardly associated with gender equality, Cochrane doesn't see it. "I thought it was pretty amazing that the world is shifting so much that this kind of message could be in Vegas."

Bliss Dance's journey exemplifies the spread of Burning Man's shock-and-awe aesthetic. Which makes sense, considering that it started with Larry Harvey's landscape business designing surreal gardens to look like outer space or lining driveways with golden bowling balls.

He and his crowd of latte carpenters were blue-collar guys with an intellectual flair, but not artists in the usual sense of the term. One of these, fellow autodidact Jerry James, who received Harvey's famous telephone call to burn something on the beach, was an actual carpenter. Since it wasn't so different from other outings they had taken with their sons, he went along. The improbability of what they launched, particularly in terms of an art revolution, was noted by an author who knew them both. "These two hesitant, alienated men soon began, unwittingly, a ritual that helped both focus and define the cultural energies that would characterize San Francisco's next decade."[1]

The ritual part came later; at the outset the scene was pure creative indulgence. The therapeutic benefits from hammering away counted much more than the eventual artifact. Not for three more years would this object—cross-beams supporting a trellis that was lying around from Harvey's gardening business—even be called a man. Its identity was incidental.

If, indeed, Burning Man was destined to harness San Francisco's cultural energies, it did so inadvertently. In the Burning Man lexicon, "aesthetic" translates a whole lot differently from how the word is conventionally used. It's much more populist than refined, and applied to work created by a whole ensemble of amateur craftsmen, most of whom wouldn't consider themselves artists. Even the playa masterpieces, the chefs d'oeuvre, rarely bear a conspicuous stamp of authorship. New York galleries and History of Art courses celebrate individual genius, the *auteur*; Burning Man eschews the cult of personality for the crew.

Take, for example, the famous red flamingo sculpture in the courtyard of Federal Plaza in Chicago. It's a local icon, always referred to as "the Alexander Calder." But consider the author of a comparable landmark piece at Burning Man—say, the fifty-foot-tall Space Whale of 2016, positioned at the head of the promenade to the Man: "the keyhole," the most coveted display space on playa. To have your art placed there is a great honor. But to the throngs who passed it daily and were dazzled by its grandeur, the artist was pretty much anonymous (for the record, it's the Pier Group Collective).

There is a designation, lead artist, that calls attention to the originator and choreographer of a piece. To be sure, some lead artists are stars. But, for the most part, they play the role of prime mover rather than singular genius. The whole cadre responsible for bringing installations to life gets the fame, collectives with names like the Iron Monkeys or the Department of Ontological Theater or Five Ton Crane.

Ever since Harvey's ragtag gang hoisted those first effigies up from the sand, Burning Man art has been inconceivable without troops to do the hoisting. Members of the teams include woodworkers, welders, glass fitters, pipe benders, lighting and sound people. What's remarkable (and another characteristic that makes this a unique genre) is that most of these workers are volunteers, amateurs learning the craft on the job. Gifting their labor. In this, though, they're considerably more than helpers. They're co-creators, often with permission to add details or elements, even to retool concepts. They're initiates in the culture.

The distinctive quality to Burning Man art, that aspect that demands its own name as a landmark movement, is this emphasis on communality. The creative collaboration among the makers extends into another previously unexplored dimension, collaboration between object and audience.

Burning Man art is meant to be climbed, manipulated, switched on and off, walked through, ridden upon. Even used for shelter. Objects are not sacrosanct, there are no DO NOT TOUCH signs. Things are conceived to be interacted with. If religious veneration defines the Renaissance, playa art is intended for play, both mental and physical. This is art with a distinct social function, meant to draw an audience. That's not a throwaway line. In its scale and activity, the aesthetic here is to function like a magnet to bring people together. It's a tool for community-building.

Given the role of ongoing participation, the importance of "incompletion" as part of the aesthetic makes sense. Inherent in the Burning Man notion of beauty is plasticity and adaptability, even the capacity to change appearance. Many installations look so dramatically altered at night when illuminated that, in contrast to daytime, they might be two distinct pieces. Others change form as they're being interacted with.

This is an aesthetic of continuous reinvention. The "shock of the new" was the revolution of Modernism; the revolution of Burning Man is "spontaneous combustion"—both conceptual explosions and literal ones. Much of the art goes up in flames (in the early days, pieces were burned because bringing them home was too much trouble). The transience of installations lends an added sense of immediacy. "My experience of Burning Man," Marco Cochrane told me, "is that it's the cutting-edge art scene in the world. Nothing comes close. We're making art for each other in the moment."

There's a second aspect of spontaneity that has nothing to do with fire. When the Burning Man Arts Foundation awards grants, a top priority is interactivity. "Interactive work convenes society around itself," proclaims the application. In keeping with the cultural imperative to dissolve artificial barriers, playa art functions to lessen distance between spectators themselves. "It generates roles...provokes actions...directs attention to the surrounding world."[2] The standard of beauty here is not "sweetness and light" (think Matthew Arnold). Praising "enlightened reason" as the criteria of greatness is a way to reaffirm the received truths of a comfortable elite. Contrary to the core, ironic and subversive, the Burning Man aesthetic doesn't pretend to moral instruction. The ultimate accolade on playa is for a piece to trigger communal afterglow *à la* Harvey's boyhood cornfield maze.

When it comes to format, anything goes. The crazier, the better, as demonstrated by the fabled Flaming Lotus Girls, six women (with two men) who came together in 2000 to make Burning Man–worthy large-scale fire art. Among them was Rebecca Anders, who calls herself a "thing-maker." Anders's parents, college professors, sent her to Cornell. As she says, they wanted her to be respectable. She showed them, though, and studied art. Her inspiration was the visionary Nam June Paik, whose work she describes as "crazy assemblies, mashing technologies into form, forcing sculpture into interaction and robots." Another was a German woman with an international reputation for mechanized sculptures that combine eroticism and violence. With such mentors, she quickly realized her destiny wasn't to show in Manhattan galleries. "I wanted to be in a place where I could blow some shit off," which, in the mid-1990s, meant the Bay Area. Home of Survival Research Laboratories, "the Big Daddy of dangerous machine art" she calls them, renowned for performance installations with titles like *An Explosion of Ungovernable Rage* and *Ghostly Scenes of Infernal Desecration.*

Anders is a recognized pillar of the Burning Man art crowd, so I crossed the Bay on BART to interview her in her studio space. It was located at American Steel, six acres of a former factory in gritty West Oakland, converted to affordable workshops large enough for bridge cranes and drive-through trucks. The neighborhood was gentrifying, though. Developers had recently bought the building. Rumor was they'd soon raze it. The makers of the oversized playa-bound constructions that surrounded us as we slumped on a lumpy sofa drinking beer understood they were living on borrowed time. This particular summer, Rebecca wasn't bringing a work to Burning Man, as she had done for nearly twenty consecutive years. Instead, there were paid commissions to finish on deadline. She couldn't afford the letdown that follows each Burn. "Afterward you're physically, emotionally, and financially exhausted."

The Hand of God, a twelve-foot woman's hand that shot soaring flames from all five fingers, put the Lotus Girls on the map. It was constructed for the 2003 Burn, in the wake of the 2002 artists' insurrection, and received the largest honorarium ever awarded at the time: $23,000. Which didn't cover materials. They were a small group then, operating from a workshop originally built by the Hells Angels, where they learned to weld, to craft with steel and copper, to use plasma cutters. And to work with fire, tutored by Dave X, the self-taught pyrotechnics expert who would become fire safety manager of Black Rock City, in charge of all things burning. Volunteers kept joining the Lotus Girls over the years (the crew numbered more than one hundred when we spoke). "I've got this *thing* about making big art," says Anders. "Larger than yourself in every regard with many humans required to make it." That's a sensible attitude when the created object is burned and disappears forever (which was not the case with *The Hand of God*). "What you have to show for yourself is the community and network of connections."

Few Burning Man masterpieces better exemplify the principle of bigness as social collaboration than Peter Hudson's *Charon*, named after the mythological ferryman of Hades who carries souls across the river Styx. Humble and self-effacing, Hudson calls himself blue-collar. "My work ethic borders on the pathological." He was a stage carpenter designing sets and props at the San Francisco Opera when, after his first trip to Burning Man in 2000, he became obsessed with building something on playa. Much like Marco Cochrane, he wanted to go large.

Kinetic sculpture became his thing: gigantic zoetropes modeled after the spinning cylinders that were the precursors to motion pictures. "So much about this art form interested me as a child. I was into magic, illusion, life-casting sculpture." He updated the format in twenty-first-century idiom, using strobe lights to create the illusion of animation. The first piece was images of monkeys swinging from branch to branch for 2007's Evolution theme. Another year, he depicted jumping figures contorting as if on a trampoline.

Rites of Passage, the 2008 theme, led to what's considered Hudson's masterpiece. A great friend and benefactor had died a few months earlier, and his passing prompted Hudson, who was raised Catholic, to reflect on the last rites. Out of that came his monumental *Charon*. It is a thirty-foot-tall Ferris wheel contained in a Gothic arch, around which rotate sculpted skeleton oarsmen posed in a boat, paddling departed souls across the river Styx. There are twenty such individual boatmen in incrementally different postures choreographed (by a Pixar animator) to flow fluidly. The steel frames, multiple pulleys, and interlocking gears are an engineering tour de force.

Impressive as it is as an object, what makes *Charon* a signature example of the Burning Man aesthetic is how it fuses with the audience. Six ropes dangle from the frame, to be tugged back and forth in pairs. Unless a handful of randomly assembled people at the base pull together with sufficient force, the wheel doesn't spin fast enough to produce the illusion of movement. Strobes are timed to flash when a brisk speed is reached, lighting each figure for an instant as it passes a fixed point. The effect is what Hudson calls "the world's shortest film in 3-D animation," an animated image of a sepulchral figure endlessly rowing toward destiny. Building up enough torque to play this film requires strenuous collective energy. The effect becomes visible, the awe factor kicks in, only through group effort. This masterful alignment with Burning Man's interactive aesthetic is the true brilliance of *Charon*.

The participatory imperative is interwoven with the principle of de-commodification. When eminent critic Robert Hughes bemoans how money has hijacked art, "the growing, tyrannous power of the market itself which has so hugely distorted nearly everyone's relationship with aesthetics," he's not talking about Black Rock City.[3] Since half the stuff gets torched, there's no real aftermarket. The metric of success resides in a piece's magnetic powers of attraction in the moment. This is what captured the fancy of San Francisco's avant-garde beginning with Harvey's first effigy. Art not as commodity nor, for that matter, even art for art's sake. Rather, art as the seedbed for a communal happening.

By year three, Harvey insisted on a four-story structure for just this reason. Others resisted. Opponents warned he was overreaching, that it was futile to try to smuggle a mega-Man onto the beach. "I remember feeling very much put-upon, feeling so disappointed that they couldn't see that it *had* to be," he recalled. "It just had to be." The taller the Man, he reasoned, the greater the community of people needed to hoist him aloft. Sure enough, the police did kick them out of town.

In the early desert years, the size of the Man was limited only by the social coordination needed to get him up. There was danger that he might topple over onto the dozens of folks tugging on the towlines. Harvey would watch from the sidelines, marveling at this audacious act of trust amongst strangers. "This is what I'm talking about!" he probably said to himself. "Solidarity and engagement." The Man *had* to be big enough to challenge the collective nerve of kindred spirits. To test their determination and courage. And, afterward, as they basked in their collective accomplishment, to sow more seeds of fellowship.

Camaraderie at Burning Man is at the heart of everything. This would become another aesthetic foundation. The subject of art pieces is limitless (tilting lighthouses or a waving grizzly bear composed of 170,000 copper pennies), and beautiful vs. ugly doesn't figure in the dialogue. The central ideal is that, when every cylinder is firing, playa art offers transport into a mind-space where conventional reality loses hold. Fantasy becomes plausible in the company of fellow believers. As the aesthetic for a cultural movement, the objective of this art is to open up unexplored psychic territory.

By 1998 Burning Man art got an official exhibition in San Francisco. A street procession in playa regalia marched to United Nations Plaza at the Civic Center, where that year's Man stood on display. Larry Harvey used the occasion to describe the Burning Man aesthetic as civic duty, "a revival of art's culture-bearing and connective function."

The next decades witnessed such things as "enormous steel climbing structures with an otherworldly ethereality," "walking neurons, dendrites drifting and gliding," a 23-foot-high steel coyote in whose stomach people could lounge, a steel and wrought-iron Zen garden, a solar-powered field of neon sunflowers, a neo-Baroque chapel covered inside and out with family photographs, a traditional Chinese temple to honor the fictional Goddess of the Empty Sea crowned by a 30-foot-tall lotus flower that glowed through the night in changing colors.

There may be pieces like these, Big Art for want of a more descriptive name, elsewhere in the world, but Burning Man is ground zero for such experiments in industrial light and magic. These displays of unbridled imagination are hugely popular. Just as Harvey had thought, scale and drama produce love at first sight. Installations become destinations, hangouts, meet-ups. The sticky glue of social interaction. People who might never enter a museum, who couldn't differentiate between Matisse and Mickey Mouse, they get it!

Darryl Van Rhey foresaw such power. But he doubted that

the populist appreciation on playa could trickle out into the world without help. "It's time," he urged, "for artists to spread out." He wanted "ordinary" people to redefine themselves as artists. One who did, as a result of his Burning Man experience, was Matt Schultz.

Back in 2008, Schultz's life was turning sour. After college he had gone to San Francisco to study animation, expecting to work for the likes of Pixar. Some half dozen years later, worn out from swimming upstream, he fled to Reno. "I was stuck in this bullshit psyche that you can't make any money in art." He got a job at a small advertising agency but this abruptly ended with the Great Recession. At loose ends, Schultz, who had long disdained Burning Man as "Bay Area hippies pretending to be hillbillies," decided to go.

He pitched a tent far off in walk-in camping beyond where cars can go, pretty much by himself. His first look at the playa by night made him think of a glowing ocean surrounded by pulsing fish bowls of light. Wandering around for the next few days, though, he remained an outsider. "I was insufferable. Rigid morals. Very judgmental. I was very good at chasing off allies and potential friends. Being alone was what I did."

And then on Thursday, running after kite flyers to take photographs, his Achilles tendon snapped. He collapsed in excruciating pain. Unable to walk, he hopped a mile back to his tent. That evening, convinced "there was no place for me," he hid behind a truck so nobody could see him crying. Schultz bottomed out.

Friday he went to medical services, which in those days could treat only burns, dehydration, and drug overdoses. They offered Tylenol and mismatched crutches. Back at the campsite, a newly made friend refused to let him remain despondent. For the rest of the Burn, he carried Matt around on his back to see the sights. Folks waved, Schultz smiled back. Never had he known people so caring. From his nadir of despair, he experienced the epiphany that turned his life around. Perhaps, he suddenly realized with numbing clarity, all the problems he kept encountering throughout his life were of his own making.

Back in Reno, with no medical coverage, he learned to drive with his left foot. At the Pneumatic Café, a popular vegetarian restaurant at the time, he struck up a random conversation with a fellow diner who, too, had just returned from Burning Man. One thing led to another; he and Kevin Chastain agreed to do an art project together. "It was the first time that someone else wanted to collaborate with either of us. Kevin and I were each other's first followers."

Out of that came *The Nest of Recollection*, "an overly complicated bird nest built out of thrift-store junk to represent childhood memories, old teddy bears and Tonka toys mixed in with poems of love and rejection." They installed it in deep playa, a quarter mile beyond the Temple, almost impossible to find. On a visit to the site, Matt came upon a brother and sister who told him they went there every day. It reminded them of home, messy and hodge-podge enough so that they adored it. "*The Nest* was the first time in my life that I made something that someone else loved."

The Schultz–Chastain partnership morphed into a collaborative joined by two dozen others the following year to erect a 300-foot-long wooden dock complete with bait shop, *The Pier to Nowhere*. It became an instant icon. The next year, they beached a Spanish galleon at the end of the wharf. "I realized that people really got off on the false memory of a lake that never existed." By then the Pier Crew, as they were now known, numbered almost sixty and was famous.

Schultz doubled down. The surreal pieces were fun but emotionally innocuous. "I'll always believe that the Temple is the most important cultural ritual in America. It gives freaks and weirdos like me, who don't believe in God or organized religion, a space to deal with death." Schultz submitted a proposal to build the Temple for 2014, which he called *Embrace*. It was unlike any previous temple, two seven-story heads with inside stairwells, their eyes locked in a heartfelt gaze, about to kiss. The figures were archetypal, indistinguishable in gender or identity, a consecration to pure intimacy. Alpha and Omega.

The jury was conflicted, some liked it and some didn't. "We got torn apart for making the Temple too representational. Too hetero-normative. We got attacked for making it a man and a woman, we got attacked because it wasn't a man and a woman." The deciding vote went against him. Schultz stayed angry for a long time. "It hurt to be rebuffed by your peers."

An honorarium was awarded anyway to build *Embrace*, just not as the Temple. Construction took a year; the cost exceeded the grant by $200,000. The geometry of the heads—with nothing straight, square, flat, or vertical—proved a carpentry nightmare. It tested the limits of the Generator, another Schultz inspiration that testifies to the Burning Man multiplier effect.

The Generator, where I spoke with Schultz, on the outskirts of Reno, is a 34,000-square-foot collaborative workshop offering space to artists—a shared pool of people, talent, and tools. It is itself a kind of giant art project, underwritten by a single couple whom Schultz discreetly refers to as "P and A." One might conjecture that P is Paul Bucheit, employee number 23 at Google, coiner of the company tagline "Don't Be Evil," creator of Gmail, and now a big-time tech investor. P and A fronted the start-up capital to launch the Generator and pay its operating costs. Why? Schultz proffers a theory. The couple, going through tough, tough times associated with personal tragedy, sought solace as part of the crew building the Temple of Transition. They heard the personal tales of others alongside them. Homeless kids, recovering addicts, down-and-outers who were finding renewal and purpose in the work. Constructing a Temple that gave others hope was redeeming their own lives. The enormity of these confessionals wasn't lost on P and A. "They fell in love," concludes Schultz, "with the compassionate community that gives them and others the space to be themselves." And funded his idea to further the process with the Generator.

Schultz had hoped to bring *Embrace* back from the playa, but the structure's laborious complexity made that impossible. It had to be burned. He and his crew worked throughout the night in preparation. The next morning, when he crawled out into the light, what awaited still causes him to catch his breath. Thousands of people had gathered to render final respects to this creation, which had been a star of the show. He sobbed at the sight.

Schultz was telling me this story a few months before the 2017 Burn. I was in Reno to attend the grand opening at the Nevada Museum of Art exhibition "City of Dust." This was Burning Man's largest museum show to date, replete with archival materials. Not all that long ago, the local citizenry regarded the Burners' annual descent as a plague. Today, Reno prizes its proximity.

Why the change? "People have a negative perception if they

haven't been there," Mayor Hillary Schieve tells me. We're in her seventh-floor office; its defining decorative motif is a chalkboard to-do list and a ping-pong table. She's a buoyant, vibrant forty-something who returned home to Reno to nurse her mother, started a secondhand clothes store and ended up mayor. When she talks about a "negative perception" toward Burning Man, she's talking about herself. "I saw everybody coming back dusty and I figured that was what it was about, being dusty all the time." Burning Man was so . . . yuck! Then she went there and did a one-eighty. "I couldn't believe I was on our planet!" Now she posts herself at the airport during Burn Week, the airport's busiest time all year, to welcome visitors from all over the world who "come to my city to see this amazing event."

Part of her enthusiastic embrace is admittedly promotional, an acknowledgment that Burning Man is an economic driver that leaves behind some $60 million in Reno every year. But part of this newfound kinship has to do with pride in the impact it is having on the city. Through the mayor's window I can see twelve-foot-high steel-block letters spelling out the word BELIEVE, a playa piece acquired from artist Laura Kimpton. Conspicuously visible along the popular river walk, it's a giant inspirational message evoking civic self-assurance.

There is a Playa Art Park in Reno's Casino District that displays pieces by Burning Man artists (more and more of whom reside locally). Mayor Schieve, a reformer in a town historically run by good ol' boys, sought to forestall developers from acquiring the land after demolition of a block of antiquated motels. "I kept advocating that we do something dedicated to Burning Man." Public opinion got on board; folks were eager to sample the art scene without getting dusty.

The mayor had wedged our meeting into an overscheduled day. A nervous aide keeps trying to hurry her along to the next engagement. To his chagrin, she tarried at the doorway to make one last pitch for Reno as a year-round destination for Burning Man tourism. "I have something in my backyard that no other mayor has." Why not, she asks, lend vacant land parcels around town to display large pieces? "It makes me sick to think of all that art just sitting in warehouses." How about encouraging developers to include a major piece in big projects, maybe through some kind of zoning requirement? "Burning Man creates jobs and culture, it's everything a mayor would want if your eyes are open."

The LLC had a vehicle to fund off-playa projects, not as grand as Reno but akin in spirit: the Black Rock Arts Foundation (BRAF). Larry Harvey, who had long criticized support to the arts as benefiting elite institutions rather than the general public, capitalized the foundation with a $30,000 loan in 2001. "It's one thing to delight ordinary people with your work," he had written a few years earlier, "quite another to 'delight' a board of directors." BRAF's mandate is to support the social function of art so dear to Harvey's heart with projects that can stimulate "immediate interactions that connect people to one another in a larger communal context."

BRAF was working mostly around San Francisco, hooking up local organizations with sculptures, when Tomas McCabe became the executive director in 2008. A filmmaker shuttling between New York and LA, he had come to San Francisco to shoot a documentary. One thing led to another, he went to Burning Man, and started romancing the woman who is now his wife, Rae Richman, at the time a consultant to Bay Area nonprofits for the Rockefeller Philanthropy Advisors and deeply plugged into Burning Man.

Under McCabe's direction, BRAF "started looking at different ways to support projects and people besides fiscally sponsoring big installations." The 2012 pilot was a $75,000 grant to hardscrabble Fernley, Nevada, to display an existing piece of Burning Man sculpture—and, more tellingly, commission a second piece to be created by local residents. The program, called Big Art for Small Towns, didn't last, "it was hard to find money," but it pointed in a direction the Project would like to pursue.

Spreading artists out into the people remains on the agenda. Few doubt that communal art—its creation, its enjoyment—is Burning Man's optimal way to spread the culture. The Project has taken up the mantle, consolidating support for on-playa and off-playa installations into a single entity, Burning Man Arts. In its first several years, around 150 projects were supported to the tune of $2.5 million. "This change breaks down the barriers," proclaimed Harvey in announcing Burning Man Arts. "Art for the playa and art for the world will be one and the same."

Conflating all venues into a single administrative unit, though, was only part of the grand design. Burning Man Arts itself was cojoined into a larger, even more ambitious domain. The creative tsunami on playa wasn't going to dry up. "The thrill of the first sight of art that should not exist," in author Jennifer Raiser's deft phrase, would continue unabated. What has changed is how the care and feeding of artists fits into the long-range gameplan. Burning Man Arts is now lumped into an inclusive unit devoted to Civic Engagement—along with Burners Without Borders, which began life providing disaster relief along the Gulf Coast in the aftermath of Hurricane Katrina, and the Regional network. The Project's organizational flow-chart, combining art with engagement, suggests the long-range tactical plan.

Leading this effort fell to a community newcomer, Kim Cook, who describes herself professionally as a "career change agent." Her virgin trek to Black Rock City came only months before her appointment as director of art and civic engagement. Among tight-knit insiders, she started out as an unknown.

Kim grew up straddled between an intellectual mother drawn to the counterculture and a religiously conservative father. She was raised in Berkeley, a flower child comfortable in ethnic scenes, and a dancer. With an MA in Art and Consciousness, her career unfolded in nonprofits: artistic director of a theater, executive director of a youth chorus, CEO of the Arts Council of New Orleans, four years at the Nonprofit Finance Fund, whose mandate is to generate opportunity in underserved communities. She learned that the best way to promote arts is by "meeting people where they are and be equipped for the conversation." Which, when talking to funders, meant being conversant with balance sheets, value propositions, risk tolerance. "You have to think about the dynamics of your business, what changes you need to make for something to become sustainable, what to sub-sidize." That was the mind-set she brought to the Project.

In New Orleans she tested theories about cross-sector col-laboration, about integrating art strategies into community transformation, about using small acts to leverage large changes. In coming to the Burning Man Project, she arrived at a place that was eager to hyper-scale along those same lines. "They're not hiring me," she told me early on, "to get art to playa. They've been doing that for thirty years."

In one of our conversations, she likened Burning Man to a mature business that needed to renew itself. "The system wants to see itself," she said in another, which I took to mean that the Project was making more unrelated decisions than was optimal. Managers were getting in their own way. The left hand didn't always know what the right hand was doing. Some found her language off-putting, a bit too corporate business-y, but she was confident that would change. "I build shared vocabularies," she told me not long after she was appointed. "I'm bilingual in hip-hop and classical culture, finance and arts."

Successes in her first years—strong art showings on playa, a well-received museum show at the Hermitage in Roanoke, Virginia, and then the mega-hit at the Smithsonian's Renwick— quieted concerns. To the people leading the Project, Kim Cook's edgy professionalism was an arrow in the quiver that had previously been missing, an invaluable addition in the quest to spread the culture. It is one thing to change the world through organic evolution, another to pursue it consciously by design. Art, as the basis for civic engagement, becomes the foundational piece in a strategy of expansion.

8

Re-Sourcing Civic Engagement

Ideas and experience together can become scalable solutions for energy, habitat, water, food, infrastructure, transportation, communications, ecology, you name it.

Black Rock Labs

The moment when the nonprofit Project replaced Black Rock City LLC marked a steep inflection in the arc of Burning Man's curve. Henceforth, and newly, the Project was committed to existing beyond the Founders. They were granted lifetime board seats, but the institution now had a destiny that would outlive them. Positioning the institution to outlast the founding generation extended the prospects for a grander scale of engagement. Art afforded one mode of doing this, but there were other ways.

Such was the big picture. But it raised a host of tactical questions. To realize the latent potential in a mission-based organization, what was the right amount of central control? When was it excessive, when was it insufficient? In the past, the answer was simple: With some few exceptions, the default was "hands off!" Let initiatives bubble up spontaneously in a thousand points of light. Now, with more at stake in the form of fixed assets and institutional momentum, there was less wiggle room. The defining issue for the Project would be like the dilemma facing any hyper-creative start-up with limited resources: With so many good ideas percolating, how do you decide which initiatives to support?

Various trails had already been blazed. These forays, however, were not the result of advanced strategic planning. Au contraire, they resulted from perceived need followed by pitched emotional response. That's how Burning Man's preeminent outreach, Burners Without Borders, started: A gung-ho crew of playa construction types left the 2005 event to help victims of Hurricane Katrina.

Tom Price, who gets credit for giving Burners Without Borders its name, recollects how events unfolded. He was a member of the LLC's external relations team that year, the folks who tour VIPs and government officials around the playa. Black Rock City in those days was a radio-silent zone, no Internet or cell phones. A total news blackout. While escorting an aide to the Governor of Nevada, Price happened upon the head of the Bureau of Land Management law enforcement team, who delivered stunning news. A massive hurricane had lashed the Gulf Coast, the BLM personnel were about to head there immediately and be prepared to stay at least a month. "That was the first I heard about Katrina."

Raised in Utah as a Mormon, Price had gone to Washington, D.C., after college to solve the world's problems. For seven years, he worked with environmental groups in the capital. "It never occurred to me that the environment needed protection until I got to a place where that was a controversial idea." Then, in 1997, he attended his virgin Burn. "It blew up my life." Not long afterward, he returned to Salt Lake City, where he became a freelance journalist reporting on climate change. All the while, he deepened his ties with the Burning Man leadership by volunteering for the media team, then the Black Rock Rangers. "I kept burrowing in more and more."

Via his Washington grapevine, Price heard about a proposed law bubbling through Congress to make the Black Rock Desert a national conservation area—which carried all sorts of implications for Burning Man. If the bill's final language was unclear whether big gatherings (i.e., Burning Man) could occur on playa, the power to arbitrarily decide would be left to the BLM. As had often been demonstrated over the years, the bureau was not always a friend. Their possessing discretionary power over Burning Man's fate would be disastrous.

Price's years on Capitol Hill had taught him how the game is played. He was able to broker meetings between LLC principals (Larry Harvey, Marian Goodell, Harley Dubois) and the bill's Congressional authors. The lobbying worked. An explicit clause ended up in the Black Rock Canyon–High Rock Canyon Act of 2000 to the effect that nothing in the law was intended to preclude permits for large-scale recreational events, a round-about way of saying Burning Man.

The BLM, however, continued to exercise responsibility for enforcing the act's mandate of protecting the landscape. Should a user of the site fail to respect its integrity, there were grounds for denying a future permit. A not-so-subtle threat.

In those days, Burners weren't quite so particular about cleaning up after themselves. Stuff you didn't want to cart home, bulky old couches and the like, would be bulldozed into a heap and lit afire, leaving behind scorched earth. Under the new stipulation, that was a hanging offense. Harvey, ever the quick political study, devised a way to ensure that change could happen without alienating his über-libertarian constituency. The solution was to subtly retool the community's consciousness with a touch so light as to be unfelt. Very quickly, a central tenet in the Burning Man lexicon became respect for the pristine purity of the setting. When the Ten Principles were formalized two years later, the eighth fundamental value would be leave no trace.

By then, Tom Price was in tight with the organizers. He was a classic poster boy for Burning Man "do-ocracy," the prioritization of capability over rank. Act first, seek permission afterward. Price had proved himself a consummate can-do guy. As news of Katrina spread through town, he hurried back to First Camp with an idea to collect supplies that could be sent along with the departing BLM officers.

The community, though, was already ahead of its leaders (more do-ocracy). On the spur of the moment, a young kid ("I think his name was Jacob") taped a big red cross to a 10-gallon water bucket, stationed himself on the main exodus road, and began taking donations for Katrina victims. Despite good intentions, money changing hands was a stark violation of Burning Man's credo that Black Rock City be a cash-free zone. "We heard what was going on," recalls Price, "and decided we had to put walls around it." Decisions were made. Media Mecca, the check-in for journalists at center camp, was designated the "official" repository for contributions in cash or kind. In one day $24,000 was raised; "the physical volume of that much cash is heavy to carry." Six tons of food were donated. "All of this was being done on the fly." Then word came that a gang from DPW and the Temple crew were leaving, entirely on their own volition, to offer their expertise in Mississippi.

"That was the first time civic engagement became physically manifested," realized Harley Dubois. "The first time we knew what it meant to be 'out in the world.'" Deploying the creativity of the community in real-life situations was a subject often discussed among the leadership but yet to be tried. "When I talk to audiences about the development of Burning Man," says Dubois, "I always say we didn't have milestones and plans. We weren't corporate in how we operated." If something was meant to be, it just sort of . . . happened. The leaders believed in libertarian populism. A cadre heading to Mississippi was exactly the kind of action long envisioned: civic engagement occurring without premeditation. "It was happening of its own nature on such a grassroots level with such good will," she recalls. "We just loved it."

Tom Price headed to Biloxi, sent by the LLC as an informal liaison. He found a big geodesic dome still covered in playa dust when he arrived, erected at what had been a Buddhist temple before the hurricane. For over a decade, the local Vietnamese community had been raising money to construct it. On the literal eve of Katrina, the day before, they held their opening ceremony. Talk about serendipity for a crew of playa Temple builders! This site became Burner headquarters.

"We started giving away food," recalls Price. "A few days later a guy showed up from Kansas City with a truck full of canned goods, he asked, 'Can you take this stuff?' We said sure." Other deliveries followed and the "temple people," as the Burners' contingent became known, established the local distribution center for the region, a 24-hour free supermarket. In his dispatch from the site, Price described how half the team "works from sunrise until long after dark, unloading, sorting and giving away food, medical supplies, clothing, diapers. Everything a person needs to survive." At the same time, the other half was rebuilding the temple, "pretty much from the ground up."

Some three hundred volunteers, not all Burners, cycled through the operation over nine months. They coalesced around familiar playa values like self-reliance and inclusion. Headquarters sent down cash infusions from the money collected at the Burn and lent support; Price's blog appeared on the Burning Man site. But there was no official relationship with the LLC. Price understood why not. "Marian and company were battle-hardened stewards of a very fragile and magical thing, they were cautious about extending the chain of responsibility. But it was obvious to them that something was being created in the moment."

A Hollywood producer donated $50,000 with no strings attached, forwarding it directly to Price personally because the operation had neither its own bank account nor nonprofit tax status. A hardworking couple caught Price's notice; their faces were familiar but in the mix of the moment he couldn't place them. "After a few days the dime dropped. It was Ryan Gosling and Rachel McAdams hiding out incognito after their recent movie success with *The Notebook*." Locals participated in playa rituals, joining in firepits and ad hoc art projects that sprang up straight out of Black Rock City. An encampment of do-gooders was fusing into a grassroots organization, citizens under their own directive doing work that needed to be done.

In Pearlington, Mississippi, where storm surges overtopped giant oak trees, old ladies were trying to haul away the shambles of their former homes because FEMA wouldn't provide a trailer until there was a space clean enough for a foundation. The Temple people took up the cause, carting off residential debris. Word spread through the grapevine and a manufacturer's rep from Daewoo appeared out of nowhere, asking what they needed. "We pointed at the catalogue," remembers Price, "and said 'one of these, one of those.'" Several days later an excavator and fifteen-ton tractor arrived, brand-new with plastic wrapping still on the seats, worth a quarter of a million dollars but gifted for the duration ("they told us to send back the keys when we were done"). The Burners provided heavy-equipment service for free, while private contractors were charging thousands of dollars.

The Temple people got included in coordinating meetings with official delegations like the Red Cross and Southern Baptist Disaster Relief. "We were the only group that didn't have T-shirts." Price was always being asked who they were. Weary of long explanations, one day he answered without any forethought. "It just came to me." If not entirely original, his response had staying power: Burners Without Borders.

In the ensuing decade, Burners Without Borders (BwB)

would spread worldwide, a loose confederation of activist pop-ups with playa experience. They were intertwined in spirit but legally independent and separate from the LLC. "I had for years wanted to see expressions of the Burning Man culture in the real world, of people becoming comfortable doing things way outside the lines," recalls Tom Price. In the past, there had been off-playa gatherings, art projects, and decompression parties to ease the post-traumatic stress of returning to civilization from Black Rock City. "But that to me was only part of it." He was seeking something with more impact, with lasting effects. Burners Without Borders proved just what the doctor ordered.

The Katrina experience, celebrated within the Burning Man community, established a sexy precedent. Later came Peru's massive 8.0 earthquakes, which mobilized a dispatch of Burn-seasoned volunteers to Pisco. "Here, like on playa, improvisation is necessity," wrote one volunteer who heeded the call. "There is no Home Depot, so even if there was money, there is no easy way to get what you think you need. Instead, you make do, and finding solutions without having what you thought you absolutely had to have back home is deeply satisfying."[4] A crew converted used vegetable oil into biodiesel fuel. Volunteers worked on irrigation canals and water purification projects, poured a new cement foundation where the cathedral had stood, built schools. The work lasted six years. By the end, few involved had ever even heard of Burning Man. "We have a lot of people who don't have any idea," said Carmen Mauk, the chief of Pisco Sin Fronteras (the Spanish spin on Burners Without Borders). "We've jumped beyond the Burner community."

Engagements bearing the BwB imprint started occurring elsewhere. Botswana, Guatemala, four U.S. cities. Each group was fiercely independent; the nature of the beast made for one-man and one-woman bands determined to run their own shows. Still, the movement began to take on the semblance of an organization with support from the LLC. Carmen Mauk was at the head, paid by the LLC. Finances ran through the LLC's bank account. That arrangement began feeling dicey to the Founders—they were legally exposed with no formal oversight. BwB took all kinds of risks, often unknowingly. Like going to Haiti after its earthquake, flying in on a private plane without official clearance. Or sending money to foreign countries, a potential legal problem with the IRS. "We encouraged them to professionalize so we wouldn't get thrown in jail," remembers Harley Dubois. "Our asses were on the line."

After steady prodding from headquarters, Burners Without Borders incorporated as an autonomous nonprofit run by its own three-person board (although the sole full-time employee, the executive director, continued receiving a salary from the LLC).

Rather than arriving on-site with a fixed agenda, BwB crews operate straight from the playa playbook. Unlike most NGOs, they take direction from the expressed needs of the community they come to serve. They listen to the people, people typically ignored by their own authorities, with the intention of preparing that community to eventually take over the mission. "Burners Without Borders really shines in places where there is not a lot of infrastructure," observed Carmen Mauk. "We're really trying to figure out, in the face of a government or a system that doesn't want to help you after an earthquake, what can we do?"

The program broadened beyond emergencies. BwB started sponsoring an annual competition for civic projects, awarding micro-grants ($1,000 max) that were minuscule by NGO standards but game-changers for the recipients. Winners included an indigenous seed library in the Philippines to preserve local biodiversity, a village program in Nicaragua combining dance with English lessons, art fairs in the Balkans that offered nonviolent-communication workshops in former war zones, and sustainable toilets at a rural children's center in Mexico. The goal was to sponsor things that others could replicate. Or, as Mauk describes the subtext, "So they can be like, 'Hey I didn't realize you could take over a crummy bathroom in a park and change it to an art gallery.'"

Indiana Jones types continued to take off to the ends of the earth, but others sought engagement closer to home. Like in Detroit, home of Doxie Katz. In 2008, when she was working as a librarian in the archives of the *Detroit News*, "all I saw was history of the city I love." The downfall of the Rust Belt had left the Detroit of her time hollowed out and collapsing. "On my commute to work, I started slowing up and noticing people living along the highway viaducts." She liked to drive around town, getting lost just for fun; on her rides she started seeing more and more destitute people. "This is stupid," she thought to herself. One day, she threw a bundle of her dad's old clothes, cans of food, some towels ("a bunch of shit") in the back of her jeep, unsure what to do next. But she had started going to Burning Man two years earlier, was en route to becoming the event's manager of open fires ("If you're going to burn anything that's not the Temple or the Man, I'm your liaison."), and truly believed in the viral power of gifting.

At an intersection where two highways converge, she spotted a homeless man and stopped. Traffic was flying by. "I don't know if I was brave or stupid." She asked if he needed anything.

"We were both terrified." Eventually he answered: "Shoes." She had a pair, stuck them in a bag with other stuff and handed it over. "He hugged me."

The Detroit chapter of Burners Without Borders sprang from that episode. Her solo travels among folks she came to call the Highway Men prompted others to donate stuff. A friend hooked her up with ten backpacks, and a program gifting survival kits was born, the Homeless Backpack Project. Money got raised. The operation outgrew Doxie's basement. A business owner offered the company vault as a central warehouse to store cans of high-calorie food (meat, beans, fish), hats and gloves, baby wipes, rain ponchos, toothbrushes and toothpaste. Along with pairs and pairs and pairs of socks, "they're like gold on the street." Twelve years later, upward of 150 volunteers gather on the Sunday after Thanksgiving in a donated theater space where, over the course of two four-hour shifts, they stock 500 backpacks. "So many people show up, we get it done fast." Over the next several months, the worst days of winter, volunteers take to the streets to pass out those backpacks. Whenever a challenge arose, Katz had a simple retort: "We're Burners, we can do this!"

Burners Without Borders, for all its latent energy, remained organizationally underdeveloped as the nonprofit Project came online. To exercise more clout, to punch above its weight, BwB needed additional administrative meat on its bones. At the same time, the Project was going through its own gestation process and fully realized that if its new iteration were to maximize impact, it needed Burners Without Borders fully inside the tent.

"We were just starting to realize how we're more than an arts foundation," recalls Harley Dubois. "We began to understand that at some point in the future, the civic-mindedness of

Burning Man could outweigh the art." With civic-mindedness its destiny, the Project reframed the role of art as a tactic as well as an end in itself. "We support interactive arts," she says, sketching out the dynamic, "that create community through which comes civic culture."

The logical next step in this process will be for the Project to up the ante by purposely undertaking larger-scale social action. Like what? One idea tossed around is to culture-hack the city. Retrofit metro infrastructures, legacies of nineteenth- and twentieth-century economies, into the twenty-first's hubs of shared affinity. Repurpose existing urban spaces to structure them as communal hot spots. Clean, green, mindfully laid-out places where people gather not just to work and reside and rec-reate but, more important, collectively create their own culture.

There's already been one try at this, drenched in Burning Man ambience: the Downtown Project in Las Vegas. Tony Hsieh, who used the money Microsoft paid for his Internet ad-vertising firm to start Zappos, moved the shoe-sellers to Las Ve-gas after selling it to Amazon for ten million shares. They took over the old city hall, located in a dumpy district of deteriorating buildings and deserted lots. Not long afterward, in 2012, Hsieh announced with great fanfare that he would use $350 million of his own money to not just rehabilitate 60 acres of the neigh-borhood but also to make it, in his words, "Burning Man all year-round." A co-learning, co-working community where, he predicted to *Fortune* magazine, "The magic will happen on its own." Headquarters delegated Cory Mavis, a longtime Burner who resided in Las Vegas, as its attaché to the project.

Things didn't work out. "The Downtown Project ended up being a bunch of people who were given Tony's money," Mavis told me, pissed off at how the venture played out. "They hired their friends and family, people right out of college with lots of zest and energy but no idea what they were doing." Even worse, the scene became irretrievably corporate; "Nobody here does gifting," she reported back. The Burning Man tie frazzled. After the Downtown Project cut the word "community" from its mission statement, Mavis departed. Not to be replaced.

What got built is now called Container Park, three dozen shipping containers converted into shops, restaurants, bars, and entertainment venues (rumor once was that there was always free beer to be had somewhere). The scene is less than robust, with many failed businesses and few of the predicted techno-synergy start-ups. "I felt no active excitement, spontaneity, or curiosity, but rather the growth of a scripted narrative and a correspondingly enforced restraint," wrote Leah Meisterlin, a professor at Columbia University's School of Architecture. "There's no meaningfully shared space with shared responsibilities."[5] The most lasting Burning Man legacy proved to be a fire-breathing praying mantis sculpture first seen on the playa.

This spirit of the commons, "a meaningfully shared space with shared responsibilities," defines Black Rock City (and is missing in the Container Park). Inculcating that same vitality, the spirit of the commons, is at the heart of Burners Without Borders. Christopher Breedlove learned it in his journey through BwB, from volunteer to project lead to chapter lead to regional coordinator to chair of its board: The indispensable element making for success is people giving freely of themselves.

Breedlove's activism-packed résumé belies his reasonably tender age. He had already burned out on environmentalism ("In college I was the kid who got my dorm system to institute recycling") and politics ("farting around with Socialism") when, at nineteen, he went to Burning Man. Back home in Chicago, he poured his fervor into starting Burner groups. Over the years, while working his day job as a designer and "experiential producer," he would help organize Lakes of Fire, the Great Lakes Regional, and an arts organization called BURN, the Bold Urban Renaissance Network ("never let a committee name anything"), then an active chapter of Burners Without Borders. When the moment came for BwB to form its own board, he became chairman.

The Burning Man Project, meanwhile, was in the process of putting its house in order. The LLC had been all over the place, its energy scattered with separate affiliations and relationships that couldn't really be coordinated. "We had to centralize," realized Harley Dubois, "in order to be able to effectively decentralize." The place to start was Burners Without Borders. From an administrative and financial standpoint, it appeared simpler to absorb than other programs. "But," she recognized from the start, "I knew it was going to be the hardest."

When approached about folding into the Project, the three-person BwB board expressed reservations. They were bringing considerable value to the table with a decade-plus experience in high-impact interventions. As an apparatus, they got results, able to move fast in contrast to the slow reaction speed at headquarters. The movement didn't want to be reined in. "We shared all the glories and drawbacks of being an organic beast," says Breedlove.

Two years of conversation ensued, fraught with obstacles. Some of the distrust was political (BwB's history with the LLC was complicated), some structural (BwB feared its penchant for quick prototyping would be held back by endless procedural roadblocks). "Burning Man," observes Breedlove, "is in the business of creating a container for creation, not creation itself." Burners Without Borders was all about the endgame.

"We kept hitting walls," acknowledged Dubois, who shifted her attention to bringing Black Rock Arts Foundation (also, like BwB, autonomous) into the Project. Breedlove and his colleagues put their talks on hold until after BRAF's deal. "We got to watch how it worked." What they saw was the prospect of being perceived wholly as a subsidiary of the Project, something they feared would appear unattractive to potential collaborators who didn't identify with Burning Man. To prevent this, they negotiated an agreement in writing that BwB could remain independent at least as a brand, keeping its name, website, social media, and newsletter.

"At the end of the day, we had three options," figured Breedlove. BwB could fully separate from Burning Man, but that would have been tumultuous. "We'd need to weave a narrative to the community about why we were no longer being supported by the Project." The second possibility was to shut down. The third, to join the Project, would free them from back-office paperwork (like managing IRS documents). "I very much wanted to see BwB continue and thrive," concluded Breedlove, "and believed working more closely with the Project was the best way to do that." The deed was done.

Burners Without Borders became a program of the Project. Breedlove was its manager, transitioning from the guy who loved being at the center of the storm into a more bureaucratic role. In the grand scheme of the long-term vision, BwB's energy and resources had to be more strategically directed. The idea was to eventually develop a general agenda for action, identify initiatives particularly suited to the strengths of the community, and seed projects to that end. Headquarters wasn't looking to micromanage, but the program manager could, at least in principle, herd the respective cats. That was Breedlove's job description in so many words; in numbers he stewards 26 Burners Without Borders chapters, 114 projects in 24 countries, 40,000 people on Facebook, and 6,000 on email lists.

"The reckless abandon of just throwing everything off to party in the playa for a week is exactly the opposite of what we need to be doing," he tells me when I ask how BwB fits into the Burning Man culture. Not that it isn't great to revel. "But it has to be balanced with work and the responsibility we have to each other on this earth."

He had recently returned from France, where he worked with other Burners at the Jungle outside the port of Calais, the squalid squatters' village where thousands of illegal migrants waited to sneak across the Channel into England. "Our project was to create a youth center with activities for kids." Shortly after BwB left, the camp was torn down by French authorities. A few months later, he would go to Standing Rock to join a Burner encampment building temporary infrastructure for protestors trying to block the Dakota Access Pipeline. The native Lakota hadn't fully welcomed the BwB approach, which caught Breedlove by surprise. "Traits valued at Burning Man, like self-starter leadership and the do-it-yourself mentality can be seen as 'settler mind-set.' No matter how well intentioned our actions and words, they often carry a deep-seated root of colonization."

"It's an ongoing balancing act," he concluded, pausing to underscore his personal dilemma. "Where to lead or where to facilitate." For Breedlove, intense by nature and restless to involve himself with the concern of the moment, these are not idle questions. How he straddles this conundrum will set the tone for the future of Burners Without Borders as it seeks to scale short-lived enthusiasm into sustainable influence.

I checked back in with him a year after our first conversation, to see how things were going. The legacy of independence continued to cause tension; there was pushback over legal and financial constraints imposed by the Project. To his BwB brethren, headquarters wanted to hem and haw over philosophy while the volunteers on the ground were itching to jump right in. If the Project wasn't responsive, dissenters were asking, why did they need it?

To differentiate and reenergize itself, Breedlove reckoned that Burners Without Borders needed a new narrative. One in keeping with a longer-term perspective. "There's never going to be enough government programs and nonprofits to fix what's going on. That's why it's so important to inspire ordinary people to create solutions with their neighbors. Sometimes it looks like disaster relief, but it doesn't have to." What would the revised story line look like? "A community activation platform for citizen-led initiatives."

"A community activation platform" exemplifies what, in the parlance of tech-driven social paradigms, is called human-centered design. Breedlove uses the term. Credit for the concept goes to IDEO, the legendary consultants famous for designing the original Apple mouse, a tool that's been industry-standard for decades now. Their engagements these days, which could conceivably serve as BwB prototypes, involve projects like revamping Los Angeles's antiquated voting system (their solution:

a device that's customizable for different user experiences, languages, and locations) and creating an entire network of affordable schools for developing countries (one idea: an accessible database of 18,000 custom lesson plans). "Design in the twenty-first century," proclaims an IDEO philosophy, "is not really about brilliant solo designers imposing solutions on lucky recipients. It is more about designers introducing methods that can be adopted and adapted."[6] That sure sounds Burning Man–esque.

Breedlove's boss, director of art and civic engagement Kim Cook, seasoned in the methodical pace and complex politics of organizations, understands his personal and professional dilemma. Playing the long game, staking out ever-expanding programs for direct citizen involvement, mandates a patient strategy. Directing BwB into new territory means learning how to harness the sense of urgency and proprietary ownership that has historically driven individual projects. Breedlove frames the platform he envisions as a dynamic of pulls and pushes: 70 percent from the field pulling the Project along, 30 percent from the central platform pushing out guidance and resources. Time will tell how well this ratio works.

Meanwhile, Breedlove has begun experimenting with pushes. "The power of BwB is as a prototyping and R&D space," he says. "Our job at headquarters is to put groups together and deploy or create a toolkit that helps the community assemble." The first attempt at this is an idea that came to him in Calais. He could see that the refugees there knew what they needed, they just lacked tools. From that insight, he devised the notion of a twenty-foot shipping container pre-stocked with a lending library of appropriate tools for generic projects like woodworking, metalworking, even sewing clothes. There would be extrusion-based 3-D printers to produce nuts and bolts. In order to power all this equipment off the grid, the container would include a generator and be wired for electricity. Mobile Resource Units, he calls them, easily transportable to wherever needed. Several of these "workshops in a box" have been assembled and dispatched to far-flung BwB sites for beta-testing.

A bookend to Burners Without Borders in activism is Black Rock Labs, an enterprise focused on sustainable energy. Its mandate is to brainstorm emerging tech—stuff like sensors, bio-mimicry, blockchains—for breakthrough applications in renewables, and bring those applications to the people in a cost-effective way, the simpler the better. It shares the same faith as BwB in relying on collective inspiration; its role in the cultural mix is to leverage the resources of the community in going from concept idea to mass implementation. "Manage, demonstrate, and scale."

Charged with bringing Black Rock Labs online is David Shearer, its board chairman and a driving force in shaping Burning Man's environmental consciousness. Soft-spoken and gentle in manner, he's another child of the Sixties who was lured from his native Oregon to Haight-Ashbury in its hippie prime. "I heard that siren, I was deeply attached to the sound of change." Other diversionary stops along the countercultural trail followed. "It took me a while to figure out what I wanted to do." In his mid-twenties, he found his answer: save the planet earth from self-destruction. Ten years late, he had a PhD in environmental epidemiology. Thence to a Caltech think tank. Thence to Toyota's Advanced Technology Group launching the Prius (his job was political, getting California to include hybrids and not solely electrics in its zero-emission car program).

He would subsequently start a company that he continues to direct, Full Circle Biochar. In dummy terms, it converts biomass into charcoal that farmers stick in the ground to enrich depleted soil. "Charcoal in the earth is like a coral reef in the sea," he tells me, looking for a metaphor I can grasp. "It becomes a biodiversity hot spot." At the same time, biochar addresses climate change by drawing down and repurposing atmospheric carbon dioxide, "converting it from a problem molecule into a solutions material." A growing customer segment for the product, with the legalization of marijuana, is commercial growers. He also chairs SkyTruth, an NGO that uses satellite imagery to monitor ecological problems such as depleted fisheries or oil spills.

It was during his Toyota stint, in 2004, that he first attended Burning Man. "I was, frankly, shocked by the high carbon footprint of the event." His response was to set up a virtual nonprofit, Cooling Man, to alert Burners to the environmental cost of indifference to spewing emissions from art cars, generators, and mammoth RVs. On the website was a carbon calculator, so people could compute their personal share of the event's total greenhouse gas (per person, it was double the national average). For those seeking pardon, the site also facilitated purchasing carbon-offset credits. Cooling Man drew praise. And it attracted the attention of Rod Garrett, the architect of Black Rock City, whose urban design was consciously set up to operate without cars.

Garrett invited Shearer to lunch back in Oakland. "He wanted to know what a renewable Burning Man might look like." Until Garrett died in 2011, they had many such conversations, talk about how to make Black Rock City a futuristic model for "leave no trace behind" next-generation urban tech.

"He recognized that extraction as a business model was on its way out and would be replaced by regenerative energy, not because it was the right thing to do but because it was going to be cheaper and more cost effective."

When Larry Harvey chose an environmental theme in 2007, Green Man, Shearer was invited to suggest ideas. Recycling and composting efforts were amped up. Event generators ran on biodiesel. He helped curate a World's Fair of emerging technologies at the base of the Man that showcased innovations like rocket stoves and exhaust-capturing algae farms. "The inner face of art and science is wonder," says Shearer. "We wanted to catalyze wonder for the demographic that is going to Burning Man."

The Man himself was stunningly illuminated in neon entirely driven by solar power. This was the result of Shearer's chance encounter with Bay Area financier Matt Cheney, who volunteered a mass array of panels. Repurposing that solar equipment post-Burn led to the genesis of Black Rock Solar, the antecedent to Black Rock Labs. Back then, governments were offering subsidies to stimulate alternative energy. Solar cost about $10 per watt to install, half of that being labor and profit. The state was paying $5 back to encourage usage. "We realized that if we weren't interested in making money, and we could get volunteers from Burning Man to install it, then we could build it pretty much for free." Which is exactly what happened. That first Green Man rig was set up in neighboring Gerlach, gifted to the tiny town's school where it produced annual electricity savings of $15,000.

A light bulb went on metaphorically as well as literally. Here was a grand opportunity to gift solar installations. Black Rock

Solar came out of this culture hack, organized as an independent nonprofit (helped by a $50,000 loan from the LLC). It resided dead center in the Burner sweet spot, combining volunteerism and gifting to spawn a viable social enterprise. Some 100 clean-energy projects were built in northwest Nevada, donated to schools, hospitals, homeless shelters, Native American facilities, a discovery museum, and a Boys and Girls Club. Highway 447, the strip of road running through Washoe County where many of the installations are erected, was proclaimed America's Solar Highway, with "more watts of distributed solar power per mile than anywhere in the United States."

Being Good Samaritans was part of Black Rock Solar's motivation, but Shearer et al. had their sights set on a more transformative role. Expensive equipment and installation made solar power pie in the sky for ordinary people, something only for the rich. Tom Price, whose role was now as Black Rock Solar's first executive director, set out to change the narrative by erecting highly visible installations where, according to conventional wisdom, they shouldn't be. "We wanted people to ask why an Indian rural health clinic could have solar but they couldn't afford it for their home or business?" Conspicuous demonstration projects, Price reasoned, would stimulate awareness and increase demand, which would incentivize production ramp-up and thus lower costs. The strategy of leading by example worked. "We went from no one applying for the utility rebates in Nevada to ten times more being applied for than what was available."[7]

Ironically, popularity undid Black Rock Solar. In 2016 Nevada's public utility commission gave the state's only power company permission to charge higher rates and fees to solar panel users (and apply the new rules retroactively to existing customers). This effectively killed subsidies, blowing up Black Rock Solar's business model. The good news? As Price had foreseen, manufacturing in the solar industry was dramatically scaling, significantly driving down costs.

Black Rock Solar declared victory and dissolved. The board wasn't quitting, just ready "to turn its energy toward a bright new star in the heavens," Black Rock Labs, which formally launched in the spring of 2017. "Renewables are approaching a point where they're cost effective," explained David Shearer at the announcement. A landmark report by the International Energy Agency confirmed the sea change, noting that the cost of electricity generated by a solar plant declined by 70 percent in the prior six years.[8] "We live in a moment of time where much of the work that's been done over the past forty years is coming to fruition," observes Shearer.

Black Rock Labs aims to seize the moment by aggregating the resources of the community. Their approach is part evangelical, part entrepreneurial. Search the world for best-in-class ideas around energy efficiency, water, habitat. Validate them from the perspective of both science and business. Build prototypes. Then widely publicize and distribute these technologies through the Burning Man network.

Through his experience as an innovator, Shearer appreciates the critical importance of accessing the right people at the right time. "It's often not the best technological solution that wins but the second or third best idea because it can be implemented." Black Rock Labs is betting on the Burning Man web of influence to accelerate disruption. "It's a demographic that thinks about these questions. Really smart people, several levels of deviation from the mean, thinking about cool ideas."

Where Black Rock Labs will lead remains open-ended. Unlike Burners Without Borders, it retains independence although highly aligned with Burning Man (Marian Goodell sits on the board). In describing this relationship, Shearer likens the Lab to Pluto, orbiting around the Project on the outer ring.

For the Lab's first trial project, Shearer has come full circle back to reducing Black Rock City's carbon footprint. He wants to someday power the whole operation with solar, starting out with a beta test of one block. He considers it eminently feasible in the near term. "We have a goal of taking all the playa and making it renewable." How? Maybe via giant extension cords linking generators to electricity distribution centers. "In my most ideal world, I'd like to see us have a carbon tax on every Burning Man ticket sold, and have it go right to Black Rock Labs," says Shearer. "Then we'd re-grant it back to the community to fund innovations in renewable."

And ultimately, in keeping with its commitment to open-source, make the tools and technology widely available. In effect, use Black Rock City as an experiment in a campaign to miniaturize solar power generation around the world.

9

A Culture of Opposites

People buy into the vision while they're here.

PAUL WOLPE

I came early to Burning Man in 2017 to visit Fly Ranch, which remained strictly off-limits. Entry was tightly restricted, visitors weren't encouraged. I needed steady persistence to get invited to join a tour, but only if I arrived before the event began. So, being that it was only Wednesday, I anticipated quick transit through the gate.

After the official start at 12:01 a.m. Sunday, the short trek in from State Route 447 can take many, many hours. Entering Black Rock City is a staged process. Local police hover along the shoulder of Route 447, stopping cars at will. Ranks of Burner gatekeepers check out vehicles, inspecting for contraband like firearms, spring-powered pistols, bows and arrows, hand-thrown spears, blowguns, and any blade longer than ten inches. After vehicles are again examined for stowaways trying to sneak in, after tickets are verified (counterfeits being not uncommon), after hugs and huzzahs and "welcome home," after virgins perform their obligatory roll in the sand . . . after all that, you enter Black Rock City. Hence the excruciatingly slow advance.

Four days early, though, I figured we'd breeze through the gate. Wrong! Our rented Toyota compact, with me crammed among two others in the back seat where the air-conditioning was feeble, inched forward in 100-degree heat that would continue throughout the hottest Burn in memory.

Get in, though, we inevitably do. I'm traveling with Will (now fourteen years old) and three of the four women who founded our camp (one of whom is my wife, Cait). We're here for Build Week; art projects are being assembled on playa and theme camps like ours put together. Save for a rudimentary grid marked by corner street signs (Awe, Breath, Ceremony, etc., in keeping with the year's theme of Radical Ritual), the vast terrain of Black Rock City is almost entirely empty as we seek the site we've been assigned. Within a few days, there won't be a free parking space.

The name of our camp is 3SP. What, you rightly ask, could this mean? It's shorthand for Third Space Place, a term derived from "post-colonial sociolinguistic theory," an element in the lexicon of community building. We're a serious lot! First Space is domestic, hearth and home. Second Space is work. The Third

Space is a café or library or park, even a shopping mall atrium where people hang out. Public places that serve as social anchors for people to mix and mingle. Like, say, the neighborhood pub at happy hour.

3SP is remarkably well run and drama free considering the stressful environment in which it exists. Over the next few days, a handful of us drive pylons, raise shade structures and wind screens, hook up a cistern, build a shower system, hang lights, and string them to a generator. Without squabbles. The reason we're so efficient, I'm pretty sure, is because our principals are women. They founded the camp and they continue to run it; the men among us are here largely to execute their directives. The short take of how 3SP came together in 2015 is that four type A wicked-smart high-achieving women broke a dozen of us off from a larger camp, trading size for a more intimate, convivial setting. With candlelight dinners and a keen eye for décor, the creative genius at 3SP is distinctly female.

When an article in the *Harvard Business Review* examined gender variations in leadership, it concluded that men may take more risks but women are actually bolder. The research measured boldness by such actions as challenging standard approaches, creating an atmosphere of continual improvement, getting others to go beyond what they originally thought possible, and quickly recognizing situations where change is needed.[9] The margin of difference was small, a few percentage points, but just enough perhaps to explain the superior performance of 3SP.

Larry Harvey once remarked to me, with the shrug of resignation that bonds males all over the world, that the whole atmosphere of Burning Man changed for the better once women became heavily involved. The testosterone excess of early days, blowing up stuff and target shooting from a moving truck, got tamped down. Harley Dubois, Marian Goodell, Crimson Rose, et al. infused not feminism but a civilized female sensibility into the fabric of the culture. And, truth be told, women did the lion's share of the thankless work that kept Burning Man running.

Indeed, it was Harley K. Dubois who gets credit (if little fame) for rendering the ongoing chaos of the first years into manageable order. In the aftermath of the 1996 debacle, when she vowed she was through with the whole thing, Larry Harvey persuaded her to return by saying Burning Man couldn't happen without her. It was Harley who was responsible for theme camps, the placement process, greeters at the gate, establishing information resources so people had a way to find each other and leave messages. "Rangers, fire, medical, Burning Man Information Radio, ingress and egress, running twenty-three different volunteer driven departments . . . I worked my ass off!" She was city manager before the job existed (and by title afterward).

Dualities like male/female framed my experience in 2017. I kept being struck with contrasts, not as contradictions but rather covalent bonds. Radical Ritual was my yin/yang Burn. My experiences kept leading me back to the conclusion that the genius of this culture is its power to reconcile opposites. This was the motor that drove the fabled life-changing consciousness that arises.

A chance encounter with Larry Harvey got me thinking this way.

I had just finished a playa interview with Skip Smith, a management guru engaged to consult one day a week at headquarters. Skip had driven me to Esalen, and en route a simpatico relationship took root. His corporate career had been in high-tech finance (the last gig as CFO of an electronic platform for fixed-income trading). If pushed, he'll concede his expertise is in technological forecasting. He's also a long-serving member of the board of the San Francisco Exploratorium, a hands-on public learning laboratory that has grown from a three-million-dollar to a forty-million-dollar operation during his tenure. No doubt, he's been brought into headquarters in hopes of, if not duplicating that experience, at least shedding light on how it was done. What does he do for the Project? "Help them look over the horizon."

Our get-together to discuss Burning Man's great opportunity and great challenge, improving the interconnectivity of the community, takes place in the cool of Skip's air-conditioned trailer. When I arrive, dusty from my bike ride to his camp, he opens a couple of cold beers. The Big Idea he savors is a master platform able to link Burners and their respective competencies anywhere in the world. A global roster of talent. A networked facilitator of civic engagement.

The challenge with interconnectivity, he tells me, is cost. How does the Project, with highly limited cash reserves, afford the infrastructure development of a platform? Curating content? System maintenance? Ticket sales is the only real revenue source, and that's stretched to the limit financing ongoing operations. You can't increase ticket prices for capital development, Smith explains, that would be a de facto tax on the citizens of Black Rock City for the sake of a global communication network. "In order for the Project to make a threshold leap in interconnectivity, it requires a viable economic model sitting side-by-side with the tool." Translation: We've got to find a way to make money!

Heading back to 3SP after our chat, I pass First Camp, the Founders' camp, and impulsively decide to pop in. I still feel self-conscious entering this inner sanctum uninvited. In the afternoon heat, nobody seems to be around. Then I notice, on the elevated deck that faces outward to the Man, Larry Harvey taping an interview with an English television crew. "Come to pay my respects," I say when he descends afterward.

As happened whenever we met, I'm not at first sure that he recognizes me. I once mentioned this to Harley, that he never appeared to know who I was; she said he hadn't remembered her name for three years. There's no affable "how to win friends and influence people" hyper-bonhomie in his manner. Not that he's rude. He's courteous but distant, coming across like a man unwilling to suffer fools who's deciding whether or not you're one of them.

I had come to learn, though, that this air of detachment was mostly façade. It wasn't disdain he was registering but social discomfort; he wasn't being judgmental but painfully shy. Once you pass mental muster, he becomes unpretentious and disarmingly candid. Our moment of connection occurred during our original conversation, when we discovered each of us had a deeply loved child of the same age. My daughter Katherine was finishing her obstetrics residency in Manhattan, his son, Tristan, was an artist welder in Marin County. I believe that gave me more credibility and counted more to him than my interest in writing a book about Burning Man.

For one possessed of such persuasive power within Burning Man circles, he is remarkably . . . *ordinary*. In some photos as a young man, he appears strikingly handsome. He was in his mid-sixties when I entered his circle, his dark hair had turned gray and his complexion wan. Even in his vintage wide-brimmed Stetson (his signature fashion statement, in homage to his father, who always wore one), he looked less like a dude than an overwrought Depression-era labor organizer. The real drama in Harvey's presence took place below the surface. His aspect, his aura, sizzles red-hot when he bites into an idea. He's a natural pedagogue. Conversation with him is less give-and-take than feeding a question and absorbing his open-ended response. Akin to playing the triangle in a Thelonious Monk ensemble, your role is to periodically punctuate cosmic riffs with a ping.

This afternoon, he had lots to talk about. We chatted in the shade of his battered Airstream. The Man might be the focal point of Black Rock City but Harvey's trailer was another geographical landmark, opening onto the First Camp firepit, which was ground zero for cultural conversations. He spent much of each event inside his trailer, plowing through the handful of books he brought along. Harvey had to be the only person who came to Burning Man to read.

This year's theme, Radical Ritual, was very much on his mind. He jumped right in, sounding oddly defensive, as if wanting to convert me to an unpopular cause. When first announced, the theme got a tepid reception. "Many people rolled their eyes or accused us of endorsing a poorly defined New Age spiritualism," admitted Caveat Magister (another member of the Philosophical Center team) after it was announced. Caveat often comments on controversial issues, his writing praised for "eloquence, depth, and sincerity." In this instance, he sought to defuse criticism by reframing ritual as nonreligious. Anything, he argued, any encounter, could constitute legitimate ritual. The only requirement was that the experience be sufficiently abnormal, weird, and wild enough to elicit "awe and wonder." No need for sacred overtones. The implication was hard to miss, that the sum total of everything at Burning Man itself constituted the grandest ritual of all. Not everybody was persuaded.

Choosing an art theme is always tricky. As previously mentioned, within the Project framework it continued to remain solely Harvey's domain, a privilege he guarded. Still, when he invited Stuart Mangrum to return as an organizer after his self-imposed exile following the 1996 schism, it was to help develop themes. "He felt they had become too serious," recalled Mangrum. "My role was to add levity and irreverence." Although originally opposed to theming ("What are we, the prom?"), Mangrum had come to appreciate the value a common reference gives artists as a shared starting point. I once asked Stuart if there was a formal process, a system, to fixing themes. His response was to laugh. "Hardly." Then he explained how it worked. Unsolicited suggestions always arrived, typically in the form of a cryptic code word like "breakfast" or "Vietnam" or "high school." Almost all are nonstarters. Occasionally, though, one strikes a fancy that launches Mangrum and Harvey into an intellectual duel. "The process starts with a joke, turns into an argument, and ends in détente."

Radical Ritual, however, was pure Harvey. He had been traveling in Asia the previous year and, stirred by the shrines he visited, proposed Ritual as the next theme. On this, he brooked no opposition. "I find the whole 'Burning-Man-as-religion' discussion tedious," Mangrum later confessed to me, "and potentially dangerous if we are to keep this a culturally generative movement." The two chewed over the concept. Mangrum wanted to take it in a playful direction. Along the lines of Falles, a fire festival he had witnessed in Spain, that dates back to the Middle Ages. Processionals in costume carry gigantic puppets, march through the streets, then throw block parties all over town. "A riot of parades carrying human figures going every which way" is Mangrum's description. Harvey's solemnity, however, wouldn't be deterred. He stuck to his vision of the Man encased in a hallowed sanctuary. Sacred space. In this instance, there was no common ground. Mangrum backed off.

Mangrum's point, however, was well taken. Playa theologians are unreliable guides. More than a few Burners are eager to share the secrets of the universe, which range from crystals to a diet of sauerkraut. On the other hand, some social constructions—spaces—are more likely than others to produce Caveat's "awesome weirdness." That's the basis, Harvey was telling me, for the importance he attaches to place. In the right circumstances, places serve as a matrix for ritual. Which prompted me to recount to him what had happened to me earlier that week at the *Tree of Ténéré*.

There had once been a real tree of Ténéré in the Sahara, a solitary acacia, perhaps the most isolated tree on earth since there wasn't another one around for 250 miles (the only reason it survived was because its roots bored down deeper than 100 feet to find water). It served as a landmark for caravans—until being knocked down, allegedly by a drunk truck driver, and then enshrined in the Niger National Museum.

The tale inspired Zachary Smith, a San Francisco artist, to conceive a Burning Man art project in homage to the tree. The piece was tall enough to stand out for hundreds of yards, its branches able to support dozens of climbers. He collaborated with Dutch specialists in multidisciplinary installations to embed thousands of LED lights in the leaves, so the tree at night was awash in changing colors. Alone that would have been quite spectacular. But, adhering to the principle of interactivity, there were headset sensors hooked up to the light system, which lucky visitors got to wear. These monitors registered vital signs like brain frequency and heart rate, which were translated into algorithms that produced the light show. Psycho-physiology as art.

My ritual at the *Tree of Ténéré* was Cait getting her playa name. A playa name is elusive as a butterfly. The alignment between context and personality must be perfect. Self-naming rarely works. Cait had been trying out "Iva" for several years, in tribute to the gigantic cardboard cutout of a stripper that once stood atop a D.C. burlesque house. Iva Price (get it?). But the fit was forced, too contrived.

Moments before her ceremonial naming, she and I paused in mid-playa during an afternoon bike ride, the *Tree* far off, a beckoning oasis. Another bicyclist materialized out of nowhere and pulled up beside us; he produced an industrial-size sprayer to offer a spritz. Ahhhhh, sweet bliss! He told us he had brought dozens with him to gift. Did we want to keep it? Sure, yes, thanks! Whooosh, away he vanished. "Gifting at Burning Man," I was once told, "is a kind of communal muscle memory. You remember the act if not the person."

Small clusters of people lulled beneath the shade of the *Tree* when we arrived. Cait offered to mist folks. A waiting line formed. One supremely grateful guy asked, when his turn came, "You know what you would be if you were a potato?" No. "A sweeeeeet potato." Voilà! The perfect name had found her. A ritual had ensued. By week's end, that's what people were calling her, a few even shortening it to Tater.

My story got Larry wound up talking about rituals. He lit his habitual cigarette. He was particularly keen to talk about the Man and the Temple of the Golden Spike that ensconced it. "I was *very* involved with this one." His emphatic tone asserted just how great his involvement was. Before he gets rolling, though, his media minder wrangles him back upstairs for the last interview of the day (Harvey is the public face of the Burn to visiting journalists looking for a quote). A Chinese blogger and her videographer await. He invites me to watch.

The thing she *really* wants to know, perhaps reflecting her Communist Party training, is whether individualistic freedom at Burning Man paves the slippery slope to ungovernable self-interest? Is the collective well-being of the whole undermined by the selfishness of the few? "Our counterargument is that you don't need to have a policeman," responds Harvey, effortlessly slipping into his own philosophical dialectics. Culture is the intermediary that restrains excess, provides internal monitoring, offers guidance. Change the culture first, he tells her, and behavior follows.

What makes culture change? "Experience." Now he's in his sweet spot. He riffs about how concepts like trust, collaboration, and community—abstract tenets elsewhere—are existentially tested and continually being recalibrated in Black Rock City. Alas, he sums up in conclusion, forces of commodification in the "real world" work against these experiential checks and balances, as human interaction governed by commercial exchange is stripped of emotional content.

The interview ends. He asks for my arm as we walk back down, he's unsteady on the stairs. I am suddenly aware of a physical fragility I hadn't before noticed. The line of his face is narrow, less full than I remember. He looks pallid. As if to fend off my concern, he tells me he had recently been sick but now he's on the mend.

After we settle back in, he returns to the subject of ritual. He references a book, *The Sacred and the Profane* by Mircea Eliade. It's informed his concept that Burning Man is a "non-homogenous space," a spirit-charged zone with a sanctified center. This is in stark contrast to the "neutral space" of everyday life. Like an Old Testament Jeremiah, Harvey foresees dread in the lessening respect for reason, hardening of the heart, and diminished joy in society's contemporary neutral space. The cardinal sin of commodification, of advanced capitalism, is that it breeds isolation and estrangement. Anomie isn't an unintended consequence, Harvey is telling me as his mental file cabinet of ideas spews forth, it is modernity's essential product.

For all this seeming gloom, though, Harvey is actually an optimist. The lesson he's gleaned from history is not that humanity is doomed but rather that it periodically renews itself. Like in the Renaissance. The evils of commodification, the pernicious effect of an all-inclusive cash nexus, are real. He's no Pollyanna. But the whole message of Burning Man, perhaps the theme to his entire life, is that there is hope. Human nature may be a constant: "I'm too old to think that people can be made to be better than they are," he once told me. But contexts can change. Physical settings. Cultural norms. Given the right circumstances, people can voluntarily become more . . . human . . . to one another.

It's in that mode that he abruptly switches subjects. "Have you been to the Man?" he asks. Not yet. "Ah, you should," he says almost compassionately, suggesting that I'm missing a great opportunity. With this year's Man, he has gotten exactly what he wanted, although, he concedes, "there were fights." Some of his colleagues objected to his plan to plant the Man's legs directly on the ground instead of raised on a platform. Harvey was adamant. He insisted that onlookers be able to gather at its feet. There was also resistance to placing the colossus within an exquisitely carpentered inner alcove, a veritable jewel box of joinery. And resistance to obscuring the whole thing on all four sides with a frame barrier high enough so the towering immensity of the effigy couldn't be grasped until you entered the alcove. "I've been thinking about this half my life, about how perception relates to our sense of bodily being," he tells me. How physical movement affects thought. That's why he fought so hard for his design. His plan was to induce viewers to crane their necks upward and feel, in their openmouthed gasp in the presence of the Man, a physical sensation of wonder.

In order to observe whether his plan worked, he had taken to stationing himself on the inner balcony within the shrine. Responses, he says, are "stereotypic." The room is remarkably still, almost hushed. People enter, then stare up and up and up at the massive Man, "captured by its immanence." To Harvey, a committed rationalist, onlookers are reacting to the mathematics of colossal scale. "They have to raise their heads back and that sensation of energy travels up their spine." Like the firing of Kundalini chakras. It was irrelevant whether anybody consciously got all this, if they understood the neurological basis for their feelings. What mattered was the physical sensation felt, the stirring of the sublime. "I'm convinced that people need that experience, desperately," Harvey tells me. "But without recourse to the supernatural."

He pulled out yet another cigarette. ("I can't write without the help of tobacco," he once told a British journalist. "Somehow the chemistry of tobacco got mixed up in my addictive compulsion to write.") Perhaps a random thought hit him at that instant, perhaps he was playing me a bit by invoking some hidden political agenda to win me over. Either way, he segued into the theme he had chosen for the forthcoming Burn of 2018. "Artificial intelligence," he confides sotto voce, as if an eavesdropper might overhear. I am being entrusted with a state secret. "People don't want to believe there's a downside to having a chip in your head," he says incredulously, smiling at me as a fellow co-conspirator. "Can you imagine?"

Stuart Mangrum would later tell me that when Harvey suggested "robot," he immediately said, "Yes!" It was a natural. Radical Ritual was about primal beginnings, the timeless foundational murmurings of civilization. Burn 2018 would go in the opposite direction, addressing a future where humanness itself would be up for grabs.

The theme came to him after he was interrupted once too often by a robo call. On the other end was a plausibly live woman. If he delayed too long in responding, or seemed not to understand her question, she was deftly programmed to inquire, as might a real person, whether perhaps his confusion was due to her failure to be clear. Or maybe he hadn't heard her because of a faulty headset. Never before had Harvey encountered anything so responsively interactive. "It dawned on me that algorithms are robots and I was getting called several times a day by robots that simulated human beings. Some of them in eerily convincing ways. The curmudgeon in me came out. I was fed up."

Yet another harbinger of modernity's dystopia had smacked him in the face. "Robots that watch us, track us, read our tweets and emails, listen to our phone calls," and then, the ultimate affront, "sell this information to other robots." In such a world, who will be master and who will be slave? Consciousness itself, the touchstone of Harvey's entire existence, might easily be hijacked. "Just talk to a robot for a while and see if it has the ability to relate various things to one another in a coherent way." He answers his rhetorical question, "Of course not!" No amount of data, no chip speed, no algorithms could do that. Robots might compute networked abacuses but that's not consciousness. "They don't think." Harvey pauses for an instant, preparing to play his ultimate trump card. "They have no soul."

He was already planning his Robot Man for 2018. Harvey had never skimped on spending for the Man, his personal art project. In the past, this had led to much dismay among his fellow leaders, who despaired of pinning him down to a budget. This one, though, would exponentially up the ante past an unfathomable threshold. He was talking about a massive robot able to shift its arms and legs "with the grace of a ballet dancer." A veritable yogi. But that wasn't all. It would interact with its surroundings, empowered with enough artificial intelligence to scan the crowd and choreograph moves in response to their cues. To play with them as they played with him. "I'll seduce them with the Man as a robot"—Harvey's tactic for fusing technology and biology—"but also as something living in their heart."

The enormity of Harvey's pursuit, which to the uninitiated must have sounded certifiably mad, didn't surprise me. I had come to realize that, at his core, he is a trickster. And this was to be his greatest trick. He wasn't pitting humanism against AI but rather mixing them together to see what emerged. Harvey was putting out a challenge, daring the Burning Man community of artists and makers to create ways to combine emotions like love and empathy with technology. I suspected that he was experimenting with something much bigger than playa art, however. The social engineer was thinking about how to tinker with the Ten Principles to recalibrate them for the coming age of robots.

I continued to wonder about all this after we parted. How will Burning Man culture respond to a disrupted social environment of big data and virtual reality and God-knows-what-follows? The measure of its vitality, the test of its potential to effect lasting impact, will depend on how well this culture triangulates dimensions of time, memory, and expectation. With ritual as the essential social building block, ceremonies must be able to respond to changing circumstances. They must successfully connect people who rely on virtual assistants, who ask their cell phones questions, to primal experiences of awe. To remain true to the Project's promise to stick around 100 years, the master narrative would need constant tweaking to incorporate the enormous technical transformations underway. The more I thought about this, the more I believed this was Harvey's ultimate quest. To blend together his humanist legacy with technological enchantment. Honoring the past while offering a credible way forward into the future.

Fly Ranch, I am convinced, is going to be the great testing ground for this next stage in the evolution of the culture, this fusion of soul and science. The emerging generation of Burning Man leadership will mature here, with the opportunity to consolidate three decades of accomplishment and to position those assets to ripple outward. In terms of Harvey's quandary, their challenge will be to exploit high-tech capability within the moral imperatives of the Ten Principles. Should they fail, should they tilt too far to one side or the other, the community won't fall apart, but its capacity to exert influence in the twenty-first century will be compromised.

The sole certainty about Fly Ranch on the day I set foot on the grounds, in my Build Week walkabout, was that while everybody agrees about Fly's importance, nobody yet has any idea what will happen here. I had come to get my sneak peek while the terrain remained untouched. "There is something about this country that just keeps calling you out," mused a working cowboy at a Nevada ranch not unlike Fly, "making you want to ride out farther, and deeper, until it swallows you whole, or maybe you just fall off the edge of it."[10] I understood what he meant as I looked around, overwhelmed by pristine purity in every direction. Matt Sundquist, who's in charge of Fly, had a similar experience the day he accepted the job as general manager. He'd never been on the property until then. In contrast to the noise of Black Rock City, the unrelenting boom-boom-boom, he encountered such stillness that at one point he heard pitter-patter footsteps behind him. He turned to see a baby coyote. "That was the moment I knew this was going to be a long project."

We were a tour group of ten and advised not to take pictures but instead "have a direct experience rather than intermediating with a cell phone." My take-away direct experience is the welcoming aspect of the setting. One feels nestled here. The Granite Mountains border one side, the Calico Hills on another. Silhouettes of distant ranges define the horizon. There are grassy meadows. Clumps of piñons and junipers. The presence of water—so far 90 pools have been counted—adds to the pastoralism. There are aquifers close to the surface, and even in late summer I feel spongy patches underfoot. This is a land, "a non-homogenous space" in Harvey's parlance, that has long been hospitable to life. Drawings are scratched in the rocks nearby, ancient petroglyph souvenirs from prehistoric inhabitants.

It's easy to wax romantic about Fly, but a sense of safety can prove dangerous. There is a less friendly side to the terrain. It's habitat for rattlesnakes. Venomous spiders. Mountain lions. The wetlands nurture mosquito larvae. "I have not experienced mosquitoes this thick and this large anywhere I've been before, including mangroves in Central America," wrote Lisa Schile-Beers, Burning Man's naturalist on the Fly Ranch project. She's documenting all the flora and fauna. "At certain times of the day near Fly reservoir, when they weren't aggressively biting, they swarm just above your head and create a thick layer of high-pitched wing-beating that sounds like an armada of tiny drones."

There's a seamy side to the history here, too. One previous owner, John Casey, "famously stole cattle, sabotaged windmills, stashed hookers from Reno in cattle haulers, and lived like a hermit king in a falling down homesteader's shack surrounded by ten thousand mother cows."[11]

We break apart for dispersed hiking. I end up walking with Zac Cirivello; along with Sundquist he is the other staff member on full-time Fly duty. His title is community and operations manager. A seasoned Burner, he had made his virgin trip a decade earlier, when he was working as a young cook in one of Seattle's tonier restaurants. "Guys would show up at the back door of the restaurant with fiddlehead ferns, mushrooms, stuff they had foraged for all morning. We'd turn it into great food." By the time he first came to Black Rock City, however, he had grown disillusioned with who was coming in the front door of the restaurant, "people more concerned with status and the price of the wine." As for many, his Burning Man experience prompted deep introspection. He got a clear handle on the source of his dissonance. "I was taking advantage of the things I was passionate about to do things that I wasn't passionate about." Instead of returning home, he joined a crew of newly made playa friends to live in a Northern California commune. "I was giving up a lot, years into a cooking career that was going very well. But there wasn't joy in it."

Subsequently he moved to the Bay Area and joined the Burning Man organization. He brought with him a key insight from his stay on the commune that would prepare him for his mission at Fly. "One of the common failures of permanent intentional land-based communities is stagnation." The first generation phases out, the excitement of novelty fades, residents leave, the newcomers who replace them inevitably lack the founders' fervor. Cirivello thinks a lot about how to avoid this downward cycle at Fly. "The fabric has to continually evolve." Adding facilities and programs is one way to do this, continually creating and un-creating buildings and offerings. But that's not enough to replenish the spirit of creation. "You don't need just available physical space," Zac cautions as we squeeze through a barbed-wire fence. "You also need available emotional space."

When the consortium that ponied up $6.5 million to buy Fly was being recruited, "emotional space" figured prominently in the pitch. Burning Man had never before tried to raise money of such magnitude. Chip Conley, who had been spearheading informal fundraising efforts, pledged to put in his own serious capital as the anchor donor if others would follow. With no development experience, the organization brought in Daniel Claussen to take charge, a conservation activist experienced in brokering public-private land deals. "We're sitting in the most fantastic million-person community, the most generous in the world," thought Claussen as he started. "And we're trying to buy that community a home. A crazy beautiful site, people who are grateful to Burning Man for changing their lives, a population that never had been fundraised." With all those advantages, he figured getting donations would be easy. He was wrong.

For starters, the Founders suddenly got cold feet. Even though they had long savored the prospects of owning the land, when the owner started actually shopping it they grew nervous. Claussen took this ambivalence as a generational thing. "They had been vigilantly playing defense for thirty years, making sure the wrong stupid shit didn't send their thing off the rails irrevocably." They worried that barriers set up to protect Burning Man would come down too far too fast at Fly, that "the little brother would be cooler than the big brother." Who would be in control after they disappeared? Those concerns eased as the nonprofit took hold. The emerging generation wasn't yet ready for the keys to the Cadillac, but a roadmap of continuity became visible.

The attitude of potential benefactors raised another challenge. Rich people were accustomed to getting something tangible for their charitable largesse. Recognition, a seat on the board, preferential treatment. But a prime tenet of the culture, with Larry Harvey the most vocal advocate even when courting donors, looked askance at quid pro quo. "They wanted their name on the gate or the right to name the hot springs after their company," recalls Claussen. Or even do a joint venture with Burning Man. None of which was acceptable. "There were plenty of people with money to whom we had to say 'no.'" The folks they said yes to were a cadre of Burners including a cofounder of Airbnb, the CEO of Cirque du Soleil, a vice president at Facebook, an early investor in Twitter and Snapchat, and several real estate developers. The release announcing the purchase underscored that none of them received any control over Fly nor any special privileges at Burning Man.

Ping Fu, whom I met at Esalen, is one of the donors with a stake in what happens at Fly. She and her daughter were on my tour, and we got to talking about her expectations. Most big funding campaigns, she explained, are long and arduous. This one, though, happened relatively quickly. "To raise all that money within a short period of time required passion," which, as she recalled, was stoked with lots of references to the ineffable ardor, the spiritual oneness, the mystical overdrive that the culture induces. That was how the passion got kindled, through allusions to Burning Man's triple crown of work, play, and love. "Play" would always center in Black Rock City, "work" in Gerlach, where the plan was to someday have extensive maker facilities. And, if the money could be assembled to buy this verdant oasis, Fly Ranch was to be the culture's capital of "love."

That message, Ping Fu worried, was getting lost. Fly was being presented as a completely blank canvas devoid of any preliminary objectives or aims. When asked about Fly's future, the stock response was "we don't know." Love wasn't mentioned. The planning process was to be open-ended, without predispositions or presumptions. The will of the community would determine its fate. Ping wasn't happy; she wanted some overarching guidance. A clear intention. "There needs to be a North Star toward which the various proposals gravitate." The Project leadership appeared not to agree. The "soulful" side of Harvey's duality was, she feared, in danger of being ignored.

Our walkabout goes on for an hour, ending up beneath a large shade cover beside the geyser pool. The younger members of the group, Burning Man junior staff, most of whom are on their first visit to Fly Ranch, peel off to skinny-dip. I follow. The water is hot enough to get my attention but not as scalding as I feared. A rocky terrace lines the reservoir, making it slick and slippery. I tread carefully, then dive under to dog-paddle on my knees to the geyser base where brilliant specks of calcium carbonate sparkle. Up close, the cone resembles a 10-foot-tall gemstone. One word, overworked but nonetheless spot-on, describes the scene: otherworldly.

Matt Sundquist is nearby. He's engaging in a low-key way, the kind of guy you figure for a good listener. "I was raised as a Christian Republican in Virginia on a 2,000-acre farm," he says about his background. "Homeschooled." Then to Harvard ('09), where he was elected student body president. "His passion and drive are matched by his charisma and verve," wrote the student newspaper *Crimson* in endorsing his candidacy. "Quite simply, nobody on campus knows more people than Matt Sundquist, and nearly everybody knows him." A dense, diverse trail ensued after college: charter school teacher, Fulbright scholar in Argentina, a student fellow at Harvard Law School (using statistics to study how corporations influence the Supreme Court), member of the Facebook Privacy team, cofounder of a start-up developing online analytical data visualization tools (Plotly), and finally a product manager at Change.org.

He came to his virgin Burn in 2016 out of curiosity. "I didn't even know there was an organization that ran it." During his stay on playa, the gig with Change.org unexpectedly ended ("they had to lay off 40 percent of the company"). The upside was decent severance; he had enough money to recharge his batteries as an apprentice gardener at Green Gulch, the Zen farm in Marin. Then, after some six months, he saw a posting for a job to be the general manager of Fly. The description hooked him: "It combined so many things I didn't know could be part of a single job." After lots of interviews, he got hired.

"I get consumed by whatever I do," Sundquist answers when asked how his first year has gone. "But this one's on a whole other level." Turns out there's more to Fly than anybody imagined. Like the presence of two dams, which bring with them a slew of laws and regulations. Or herds of federally protected horses. Or complicated zoning codes that limit how water can be used. Or invasive species. Every detail is complex and nuanced. "Nothing's ever taken up so much mental bandwidth."

He repeats the company line about Fly's future. Its potential agenda is as broad as the horizon. As we soak in the pool, he spins out his techie version of how he sees the operational aspects of the project unfolding. He's imagining a multi-layered platform that renders every facet of building Fly transparent, from topographical photos through to final project implementation. And all this happening at unprecedented levels of mass participation, looping in stakeholders at each stage. "Anything we undertake, we want to make sure the right people weigh in on it." What he has in mind, this comprehensive collaborative execution platform, doesn't yet exist anywhere. There are components around that can do decision-making and discussion, that post questions and answers, that manage projects and divvy up tasks. But no single integrated platform. "This interests me," says Sundquist in a colossal understatement. "It's a big part of why I'm here."

He's realistic about how high he's set the bar. Best-case scenario, it'll take five years to string together all the pieces. But he's thinking on a grand scale, envisioning a tool for community-building applicable far beyond Fly. An open-source resource that would be a game-changer for social activists reviving neighborhoods or constructing centers for displaced refugees or even creating an entire town. They would be able to efficiently engage stakeholders in design and construction. Nobody a client, everyone a participant. Sundquist is not inclined to hyperbole, but as we towel off, he speculates that this participatory platform on behalf of intentional community could end up being Fly's most consequential accomplishment.

After the peaceful tranquility of Fly, Black Rock City seems hyper-clamorous upon my return. It reminds me of Marrakech at the end of the Saharan trade route, a raucous terminus packed with snake charmers and fire eaters. The two venues, Black Rock City and Fly, so different from each other in ambience, are going to be difficult to integrate into a common state of mind.

My impromptu conversation with Larry Harvey a few days after seeing Fly, though, helped resolve the contradiction. Or at least hammer it into manageable form. Reconciling the duality he warned of—sentiments of the heart versus rational AI—will really have to be the dominant motif at Fly. For the Project's 100-year movement to prevail, it needs to be a healthy convergence of Ping's love with Sundquist's agnostic technology. Hanging in the balance is the coherence of the culture, and, with it, the prospects for that consciousness taking root in the fast-coming brave new world.

I sought out an eminent futurist, expert in such matters, to inquire if he thought Burning Man culture was up to the task of exerting consequential social influence. Paul Wolpe is a professor in the departments of medicine, pediatrics, psychiatry, and sociology at Emory University. He's also, as a side gig, a senior bioethicist for NASA, the space agency. ("Before you think about what you put on a manned craft to Mars, you're going to have to define a set of values by which you're going to make tough medical decisions.") We met atop a tricked-out double-decker sound bus at the 2017 Burn. I called him in Atlanta a while afterward, eager to hear his thoughts about whether Burning Man culture had potential to change the world.

"These kinds of experiments get generated at historical moments when people imagine and configure different kinds of responses to deeply felt challenges," Wolpe explained. As illustrations, he cites millenarian movements like the Shakers or the communes of the Sixties. "The prototypes create exemplars in hopes they might spread." The hard truth, though, is that while most movements experience temporary victory, the long-term prognosis is defeat. By historical measures, he said, Burning Man has already succeeded. The Shakers numbered 6,000 at their peak, Burning Man has 75,000 and a waiting list every year. Whether this makes for lasting social consequence, whether Burning Man will be the movement that beats the odds, remains uncertain to him.

"Its uniqueness is in the combination of ideas and emphasis," Wolpe concluded. It's the way the Ten Principles undergird an operation that, at least for a week, holds together a massive, diverse populace. "People buy into the version while they're here."

However, he's quick to add, the internalized rules of engagement won't necessarily apply elsewhere. Some norms can't travel (e.g., running around naked); some are only partially exportable, like gifting or de-commodification. But some—self-reliance, inclusion, no trace left behind—certainly could be integrated into secular society.

The problem, he suggests with professional skepticism, is that once you start cherry-picking select principles, you're tinkering with the culture's coherence.

10

In Pursuit of a Utopian Dream

A man who sees and feels deeply.

HARLEY DUBOIS

I had a trip to San Francisco planned early in 2018, and I invited Larry Harvey to join me one night for dinner. Alas, he apologized, he wouldn't be in town while I was there. Traveling abroad to raise money for Robot Man, which he figured would cost . . . who knew how much? Millions. Budgets never stood in Harvey's way when it came to the Man, but with the advent of the nonprofit, stricter financial protocols reined him in. To build the chef d'oeuvre he was contemplating—the AI robot dancing in sync with the crowd—he needed to land some big donations. Thus he was off to the United Arab Emirates, a fact that gave me pause when I imagined the same guy who carted a ragtag stick figure to the beach now calling on sheikhs for money.

That would be our last exchange.

I hadn't counted on Larry Harvey dying. Nobody had. At the Burn that summer he seemed frail, but he attributed his wan appearance to a recent illness from which he was now recovering. The news of his massive stroke on April 4 came out of nowhere and hit me like a blow to the gut.

Details were scarce. The story I heard was that his girlfriend Cheryl Edison, worried that she couldn't contact him, called his son, Tristan, who found him collapsed in his apartment. At California Pacific Medical Center, where he was rushed, emergency brain surgery was recommended. Without it, Larry would likely not survive the night. Tristan explained the situation to his seemingly incommunicado father, who lifted an arm, which Tristan took as a signal to proceed. This would be his final communication with the world.

For three weeks he hovered in a coma. A trio of family (brother Stewart and nephew Bryan along with Tristan) and Project compatriots kept constant vigil at his bedside. "We set up a 24-hour watch," Harley Dubois told me, "he was never alone." Dear friends wrote letters that were read aloud to him. "We resolutely held out for a miracle," said Marian Goodell. "If there was anyone tenacious, strong-willed, and stubborn enough to come back from this challenge, it was Larry." The miracle didn't happen. Sunday, April 28, 2018, he passed.

Obituaries ran throughout the world. Many quoted Stewart: "Larry's great gift was seeing possibilities, then possessing the eloquence to persuade others to come join in and make it all happen." Ever the supreme rationalist, Harvey didn't believe in an afterlife. In making the announcement to the Burning Man community, however, Stuart Mangrum begged to differ. "Now that he's gone, let's take the liberty of contradicting him and keep his memory alive in our hearts, our thoughts, and our actions."

Stunned disbelief filled Burning Man headquarters on Monday after Harvey's death. Little work got done; the day was spent exchanging Larry stories. A group went to Baker Beach, to render tribute to the Founder at the site of the founding. Folks kept showing up at the office. Ex-staffers, Burners, friends. A circle where he had always been at the center. There was richness to the sadness; the grief of loss was tempered by gratitude. "As you looked around at all these wonderful people," recalled one who was there, "you realized that he was responsible for bringing us together. None of us would have ever known each other if not for him."

Just days before, Harvey had savored a spectacular triumph. The Smithsonian in Washington opened "No Spectators," an exhibition of Burning Man art and culture. The Renwick Gallery, the oldest art gallery in the country that was called America's Louvre at its 1874 opening, was entirely given over to the show. A 15-foot paper arch covered in pictures, giant origami mushrooms, a 12-seat Art Deco movie theater atop a bus, the grand salon devoted to a David Best temple were all on display. The first weekend, 20,000 visitors streamed through. Crowds kept coming; many had never before visited the Renwick (or even knew it existed). Before the exhibition closed, it would attract half a million attendees. Political VIPs wanted private tours. Museum shop sales boomed with playa-esque outfits and objets d'art.

Harvey often likened himself and his friends to weeds. "A weed is not a biological category," he would frequently say, "it just means any unwanted plant." With the Renwick show, the Burning Man crew were weeds no more. The Cultural Establishment had taken note. No less a custodian of the national heritage than the Smithsonian Institute accorded recognition. Even more important, the public loved the stuff. "Showcasing these works in a major Renwick exhibition definitely marks a sea-change in Burning Man's identity," said *Artes* magazine. But, as the reviewer shrewdly observed, heightened celebrity could prove a two-edged sword. "Will these selfless craft-makers resist the siren song of commerce," she asked, "or will the groundswell of national recognition transform Burning Man into the maker movement's hottest consumer ticket?"[12]

The moment of Harvey's passing seemed unduly cruel, coming at this glorious apex. But maybe, upon further consideration, the timing of his death wasn't at all ironic. Rather, it might have been a hidden blessing from the gods. With the event now regularly showing up on bourgeois bucket lists, the genie was out of the bottle. The next generation of leadership was going to have to confront the dilemma of whether to resist the siren song of commerce or become the hottest of consumer tickets. Stay pure or go big? There were persuasive arguments on both sides.

Unlike before, however, Harvey wouldn't be in the middle of heated controversies seeking, if not the golden mean, at least a working compromise. His passing at this moment might actually offer a hidden grace. His reputation, his aura, his stature were untarnished. Perhaps, as per the mythic rites that Joseph Campbell describes, the symbolism of the moment was perfect for the final act in the hero's journey.

"No Spectators" struck the pitch-perfect coda to Harvey's life. Even with its sanitized version of the playa scene, rendered gallery-respectable so as not to offend, the stunning accomplishments of unleashed collaborative creativity were too powerful to deny. And this, more than the art or even the event itself, was what he had so arduously nurtured. On display at the Renwick was Harvey's conception, supplemented and enhanced by others, of a culture that prioritized human connections. In physical terms, Black Rock City was an expanded version of the maze he wanted to carve in his father's cornfield. A realm of hidden possibilities.

The show was jam-packed with paradoxes. But that was exactly Harvey's message: Contradictions weren't a problem but an opportunity. This was the point with the Ten Principles, they were inconsistent on purpose. The *Washington Post* critic Michael O'Sullivan prepared his readers for an anomaly of the show they'd have trouble wrapping their heads around. He cautioned them to be ready for complex ambiguity, the mixing of "the pleasures of visual hedonism—eye candy, for lack of a better word—with works of a more spiritual, even sacred, bent."[13] The Renwick exhibition resisted simple categories. Like Burning Man itself.

"No Spectators" was conceived by its curator, Nora Atkinson, to awaken the staid corridors of D.C. to the creative maelstrom emanating from the Pacific coast in general and Burning Man in particular. She came to the Renwick in 2014 from Washington State, hired for a newly endowed curatorial position specifically dedicated to craft. "I found lots of things that were common on the west coast a revelation here." Although she hadn't been to Burning Man, lots of her friends went. Mounting a Burning Man show was always on her short list, but in Seattle "it didn't make sense, a big crowd already knew about it and attended." The east coast, though, was terra incognita.

The conviction that she simply *had to* do the show came at Art Basel Miami, the ultra-chic ne plus ultra art fair that has few rivals in its "ability to generate that deep-seated biological longing to be with the cool kids."[14] Nora's take on this mother of all art markets, though, was much different. She was repelled by the "sickening high-consumption scene of commercial art." It was wholly at odds with her passion for craft as independent artisans working for their own pleasure, "something that matters to the people who make them." And what better articulation of craft than Burning Man with its gifting aesthetic of "making cool stuff for the people around you."

The Renwick had a reputation as a sleepy, obscure institution when Nora arrived. That was about to change. A full-scale, two-year renovation was underway, accompanied by a campaign to reinvent its brand as contemporary and relevant. "I'm always wanting to poke the bear," she admits, "I like to play outside the boundaries." The stars were aligned. She cautiously pitched her proposal for a show, worried that there wouldn't be much appetite in a conservative town that knew only "the trite stuff about Burning Man." Just the opposite happened. The powers that be, conceding they didn't know what to expect but were willing to take a chance, gave her the green light.

Next she had to get the Project itself on board. In May 2016 Nora introduced herself via email to Harley Dubois. Another museum show was already in the works for the following year, at the Nevada Museum of Art in Reno. Curator Ann Wolfe was drawing from the archives to mount a history of "the pivotal moments that shifted the course of the event from countercultural to global phenomenon." Nora had something even bigger in my mind. She was talking about commissioning artists to make installations. Worried that such a grand show might smack of commercialism to Burning Man folks, she framed her approach with an academic slant. "In my eyes, one of the defining elements of the modern craft movement from its British origins with William Morris and John Ruskin to the American Arts and Crafts Movement has been its pursuit of a utopian dream which emphasizes the values of handwork, self-sufficiency, inclusiveness, and community," she wrote in that initial correspondence. "Thus you can see why Burning Man is such an interest." Her interest was reciprocated.

In 2017, with the show scheduled but no roster of exhibits yet, Atkinson went to Burning Man for the first time to saturate herself in the milieu. The physical conditions didn't intimidate her. "I'm the youngest child from a family of campers." Top-echelon Project leaders served as personal guides, introducing her to a smorgasbord of art and artists. The curator in her made mental notes that would influence the exhibition. One thing that stood out was the energized relationship between object and audience, in part because of the active effort required to visit a piece. "You spot it on the playa and have to go to it, not like in a gallery. You have already invested time in the piece just getting there." The effect of environment also was important in the aesthetic experience, like the way loud music dissipates into a resonant bass that incrementally fades away as one moves off into the distance. She watched the Temple go up during construction. Naked, without people, it was nice but not mind-blowing. A week later, now "a mass of human emotions," she was too moved to enter. "It broke me in half."

The most powerful impression, the expression she wanted the Renwick show most to convey, was "boundless human creativity." Doing this indoors and funded by a federal entity hypersensitive to controversy would be hugely challenging. Nor would Burning Man easily agree to restrictions and requirements. There were lots of conversations. But Atkinson was convinced that the show would give both institutions a chance to broaden their brands. For the Smithsonian, so comfortable being historic and looking backward, here was an opportunity to get in touch with something current. And for Burning Man, a hippie-fest to the general public, it offered legitimacy.

Savor that legitimacy the Burning Man community did. When "No Spectators" opened eighteen months later, all six Founders attended. Michael Mikel, Danger Ranger, used the occasion to make his first trip to Washington in his seventy-plus years. Several stayed at the elegant Hay-Adams, a five-minute walk to the Renwick through Lafayette Park past the White House, the same hotel where Barack Obama and family resided prior to his inauguration.

There was an insiders' preview to honor the artists the night before the public reception, a friends-and-family affair with the run of the place and a luscious sushi spread. So many moms and dads were present, numerous wheelchairs with aged relatives basking in their offspring being blessed by the Smithsonian. Larry Harvey was in his element, spending quality time with people he loved, completely at ease. "He was talking and yakking with the biggest smile on his face," recalled Harley Dubois. "He was having a blast."

With the party winding down, Will Roger went outside to take the air. He was in a pensive mood, reflecting on his extraordinary journey. Fire dancers performing with the White House as their backdrop summed up the distance he and his cohorts had traveled. Of late, from his home in Gerlach, Roger had been working on a book. It was now finished, *In Search of the Common Shaman*, a metaphysical memoir in which he writes, "My paths were many . . . years of highs and lows . . . myriad successes and failures."

A small group exited the Renwick as he sat musing, Larry Harvey among them. They were about to embark on a bicycle pedicab tour to see Burning Man pieces that had been installed in the surrounding neighborhood (so unexpecting passersby could experience the serendipitous shock of Black Rock City). Harvey strolled over to Roger. "He had that expression on his face filled with wonderment and innocence that only Larry has. He looked at me and said, 'Will, you've re-invented yourself again.' Then walked away. Those were his last words to me."

The public's first crack at the show came the next evening. An opening reception packed the house at $120 a ticket (valet parking and open bar included). Wildly costumed Burners from all over were there (invited VIPs unfamiliar with the practice confessed to feeling self-conscious in "normal" attire). The Founders dressed Burner chic. Michael Mikel looked cowboy *GQ* in a long black duster coat and paisley vest, Harley Dubois wore a "fuzzy silkish robe-style jacket, very Burning Man but also crafty and sophisticated." A woman in a hoop skirt accessorized with filled champagne glasses moved through the house offering drinks. Savories beckoned at every turn, curated to resemble playa yummies with a gastronomic touch like candied bacon, luridly colored doughnuts, a tree strung with soft pretzels.

Ray Allen, the Project's thoughtful general counsel, left early. The Renwick scene was too hectic for him. He returned to the spacious Airbnb town house that he and a few other leaders had taken over and brewed himself a cup of tea. As he sat sipping, to his surprise Larry Harvey entered. He, too, quit the gala long before closing time. "I'm an introvert," says Allen, "so is he." They sat in the kitchen talking their usual idle chatter: jokes, politics, philosophy. "Larry seemed content, much more so than usual. I think it was the satisfaction of seeing what he created showcased at the museum." After a while, he retired to bed while Harvey went outdoors into the rainy night to smoke a cigarette. Harvey's air of uncharacteristic peace during their conversation assumed significance in retrospect, when Allen got the news a few days later. "I think it was the conciliation of seeing his life complete."

Stuart Mangrum didn't attend the Renwick opening; he was on the road at regional convocations starting in Colorado for the Rocky Mountain Burning Arts Summit, then off to France for the International Leadership Summit. When he heard the news of Harvey's stroke, he got on a flight back to San Francisco where he joined the bedside vigil.

The two had last been together at a meeting of the Philosophical Center, where their conversation got around to an article in an artsy magazine Stuart had recently read. It proclaimed that Burning Man had spawned a major international sculptural movement. Harvey was almost dismissive of the piece, saying Burning Man was changing something much bigger than a single genre. It was, he insisted, disrupting the entire commercial art market. The prevailing model favored a few select works that were accorded recognition and outlandish prices; he foresaw a whole different model that would owe its origins to Burning Man. There would be lots of installations everywhere, not to be owned but to be used, products of collective effort by groups of people who hadn't previously considered themselves artists. Within this new cultural paradigm, the value of a piece would be measured not by the price it fetched but rather the richness of collaboration it inspired. "We're just getting started" were Harvey's last words to Mangrum.

Most everybody from the Project cleared out of D.C. after the public reception except for Michael Mikel and his wife, Dusty. He was to address a mid-Atlantic regional meeting the next week, so he stuck around. During the interim they traveled south to St. Augustine and Savannah, Monticello and Williamsburg, making an American heritage road trip informed by his love of history.

I got to know Michael several years earlier, riding back from Esalen with him and Dusty. They live in Reno but keep a place in San Francisco. We shared a couple of tumblers of late-afternoon whiskey that day as he toured me around his extensive collection of west Texas memorabilia, vintage rifles and such, much of it family heirlooms. A self-described "outlaw kid," he planted homemade smoke bombs in school lockers. He showed talent as a DIY maker early on—at fifteen he sewed a shirt from scratch. "The boys tore me up but the girls loved it." In the Navy, as a top-secret communications officer, he watched the Gulf of

Tonkin incident unfold over the wires. This was the purported attack on an American vessel that gave Lyndon Johnson pretext to wage war in Vietnam. The portrayal of the event by the president bore little resemblance to the chatter Danger Ranger had heard. "The only thing known for certain was the spotting of what was perceived to be torpedo wake," he recounted on our ride in from Esalen. No other evidence. "I came to realize that the government tells structured lies." The experience changed him from pro-military to anti-war.

He moved to San Francisco after the service, where he had another life-changing moment reading Allen Ginsberg's epic "Howl." A landmark in the national literature, "Howl" is "a lament for the Lamb in America" and the "monster of mental consciousness that prays on that Lamb," namely industrial capitalism ("Crossbone soulless jailhouse and congress of sorrows"), as Ginsberg described it. For Mikel, the poem "exploded my world view." He hung around the periphery of the Beat movement, which led to the Cacophony Society. Which led to Baker Beach. Which led to the first desert Burn, where he dug the fabled line in the sand, on the other side of which he told the gathered multitude, "Everything will be different." Events proved him right! These days he sits on the board, a staunch defender of community standards.

At the 2017 Burn, Cait and I took to hanging out at Michael's inviting camp. It's in the quietest part of the playa, on the far side of the fence beyond which motorized vehicles are prohibited. The vista eastward to the surrounding mountains is unblocked, making for spectacular sunrises. With his vast knowledge of the terrain, he chose this spot as the specific coordinates least likely to encounter a dust storm. The camp habitat is shady and breezy, an ambience that combines *Out of Africa* rattan wicker with the corner bar of a friendly speakeasy. No surprise, from this vantage point Dusty and Michael tend to enter Black Rock City sparingly. "We let Burning Man come to us."

The spirit of this place, not of isolation but rather independence, carries over to Michael's interactions as a Project elder. In council, his has often been the voice of dissent. He was the one, for example, who filed arbitration demands to change the structure of the LLC and to protect the trademark. Not that he is necessarily opposed to his colleagues, but he's a stickler when it comes to defending the values of the movement. Call him a folksy purist. The philosophy behind the Black Rock Rangers is a quintessential example—he organized them not as constables of the law but rather guardians of the culture. While all the leaders are committed to protecting the spirit of Burning Man, few translate ideology into practicality better than Michael.

His relationship with Larry was complicated. In the beginning, they were close. When Harvey was broke, he'd pay his rent. Or leave bags of groceries at his doorstep. Sunday mornings, Michael showed up with doughnuts and they brainstormed for hours over black coffee. Six months later, he'd hear his words emerging from Larry's mouth. "I'd come up with ideas, Larry was the front man. He was a good talker." It was Michael who prompted him to don the signature Stetson. "If you want people to listen to you," he advised back in the beach days, "you should wear a hat. It lends respect. Why do you think a king wears a crown?"

The Great Split of 1996 came from deep fractures between Harvey and some of the other organizers. Mikel shared many of the concerns that provoked the schism. Over the years, the two had some nasty spats. Trying to out-reason Harvey was futile. His arguments within the inner circle were notoriously vague; rebutting them was like walking in quicksand. Mikel stopped trying. By the end, they maintained a strained relationship in their official roles; off duty they avoided each other.

From Reno, Mikel tends to a comprehensive collection of historical Burning Man materials he has scrupulously maintained, much of which has been donated to the archival collection at the Nevada Museum of Art. It was in his capacity as house historian he took to the stage in Baltimore to address the 2018 Mid-Atlantic Leadership Conference with a presentation entitled "The Five Ages of Burning Man."

The weekend assembly drew a slew of Burners from the region to a cavernous warehouse located in an area called Hampden, a district of vacated nineteenth-century flour and cotton mills. Streets of small houses rise up the surrounding hills, built for migrants from Appalachia who came to work the factories. For decades this had been a downtrodden patch of decaying urban archeology. No longer. The housing stock is being renovated. Artisanal restaurants with names like Cosima and Cypriana have sprung up. There's a local whiskey bar offering flights of single malts. And the site of Ranger's talk is now called Creative Labs, a 24,000-square-foot art incubator with studios for musicians, clothing designers, multimedia artists, photographers, painters, and even a hair stylist. It's a cousin of the incubator in Reno.

This transitioning district exemplifies the target at which Burning Man culture is drawing dead-aim. It's not just hipsters who seek places like Hampden. There's working-class folk.

They're here to recast the American Dream from an oversized consumer lifestyle (which is decreasingly affordable) to a simpler but richer celebration of community. The fruits of consumption, with signifiers like expensive automobiles and outsized McMansions, is giving way to the ideal of neighborhood. On the happiness scale, meaningful interaction and shared experience are moving up, material possessions moving down. The roadmap to this transition comes right out of the Burning Man playbook.

If not precisely in those terms, this is Michael Mikel's discussion subject. He traces the evolution of Burning Man, naming each period. He concludes with a forward look at the current phase, which he calls Scarcity. The promise of Burning Man, he argues, is at a critical threshold. All the people who want to come can't be accommodated. The historic model, with the playa as action central, has hit its ceiling; the demand for Burning Man's culture of civic design exceeds Black Rock City's supply.

The solution, were it even possible, isn't to expand attendance. Fly Ranch—"a physical polestar that our culture can revolve around all year" is Danger Ranger's description—will help. But even that's not going to be enough, he tells the crowd.

The institutional format cries out for the next evolutionary phase. The community's energy needs to shift. Mikel calls himself a futurist, that's his chosen profession. "The problem with being a futurist," he frequently observes, "is that you see things before everybody else and you have to wait for them to catch up." From his spot on the podium, he had already spotted a sea change in the energetic core of the community. The clusters of Burners worldwide, the Regionals and beyond, were destined to become the prime movers of the culture.

The nonprofit Project came about from the Founders' acceptance of their mortality, their imminent "exit on that silver spaceship." The seed was being passed to folks like those in this audience, to the leadership councils and eighty Regionals in three dozen countries. "You are responsible," Mikel exhorted them, "for forging the next link." He was confident they were up to the challenge, primed to build on the Ten Principles in ways that nobody could yet imagine. And as they do, Danger Ranger assured them, the sixth phase of Burning Man will name itself: Abundance.

A glimpse of what Mikel sees awaiting was available a few months earlier at the 2018 convention of the U.S. Conference of Mayors in Washington. The ballroom of the Capital Hilton Hotel was tightly secured for the final plenary, a session honoring public initiatives in the arts. The wife of the vice president, Karen Pence, and the minority leader of the House of Representatives, Nancy Pelosi, were making remarks. Following awards to the mayor of Dallas and the governor of Maryland, the Project's director of communications, Megan Miller, was called to the dais. In introducing her, Reno mayor Hillary Schieve declared that her city owed much of its "rise from the abyss" to Burning Man.

A "Network of Dreamers and Doers from Your Cities" was the title of Megan's comments, a primer to initiate the mayors on the abundant resources available to them close to home. "Like you, we spend a lot of time thinking about urban design, power distribution, traffic patterns, and the importance of open space." Black Rock City offered a powerful model to urban managers, a kind of experimental proof of concept: "When people build their own city, they become deeply invested in what happens there. They take care of it and of each other." She told them how Burners were building "more vibrant, inclusive, resilient communities" through experimentation, citizen-driven initiatives, and high levels of participation. Her subtle call to action came through loud and clear: Make use of us.

In a breakout session afterward, Stuart Mangrum drilled down to basics, citing various Burning Man axioms cities could put into play. The power of volunteerism. Spontaneity ("the first rule is there are no rules"). Collaborative creation ("maker spaces evolved out of Burning Man and the need to make big art"). Interactive art as a social magnet ("a draw for communal gathering").

I scanned the room as Stuart spoke, wondering whether his audience could take this agenda to heart. The bureaucratic regimen of government is at the opposite extreme from do-ocracy. When I left the room, I got an answer. A big placard stood outside, bearing the words "Before I leave office, I will . . ." with blank spaces for responses. An example of participatory art in action. One scribbled intention stopped me in my tracks: "Before I leave office, I will . . . give my state-of-the-city address at Burning Man."

Here was a marker in the sand worth noting. In this assemblage, there was a mayor who figured Burning Man has something important to teach—much like Grover Norquist, the über-conservative Republican strategist famous for the Taxpayer Protection Pledge, who wrote after his first Burn, "Some day I want to live fifty-two weeks a year in a state or city that acts like this."[15] Or, say, high-level marketers who admire Burning Man's approach to branding ("the best kind of marketing today is the kind that is used to build cultural movements").[16] Similarly it has inspired the hexa-yurt (a six-sided design that can be built from various materials, with plans available online) and Shift-pod (a nonconventional tent invented by a long-time Burner who wanted playa shelter that could be set up quickly, withstand heavy winds, and be big enough to stand up in)—both are being used as emergency shelters.[17] And the playa has infused a Paris Fashion Week designer's collection ("the dream wardrobe of a neo-tribalist futurist princess").[18] An art magazine calls Black Rock City the ultimate venue to experience modernism ("Burning Man points toward our latent capacity to reinvent our world according to different principles and ideals").[19]

The first large-scale open-source LAN network was built on playa (the size of a shoebox, it required only fifty watts of power and offered free cell phone service). As were autonomous aerial vehicle delivery drones, portable graywater filtration systems, solar-powered bicycles, a crowdsourcing digital approach to public health, and innovative techniques for environmental cleanup.

The point of this litany is that much of what occurs at Burning Man is already being taken *seriously*. As at the Renwick. "If you're trying to do something new and looking for solutions to interesting unique problems," says CEO Marian Goodell, making the point that Black Rock City is an ongoing experiment, "you come here and find it's a blank slate." What happens at Burning Man doesn't have to stay at Burning Man.

The neoliberal order of the last century is being hollowed out big-time by technology and climate change, economic and political upheaval. There're going to be voids to fill. Lots of Marian's "blank slates." Burning Man offers one compelling example of how to reorganize social behavior. Not through the usual religious zealotry or financial incentives or dictatorial edict or black magic psyops. It's happening through collaborative creativity in service to human needs. And, to repeat Larry Harvey's farewell to Stuart Mangrum, "We're only getting started."

Even in Harvey's absence, indeed perhaps *because* of his absence, the community was prepared to take up his mission. Ownership of Burning Man was vested in the nonprofit for precisely such eventualities. The Project was organized to provide continuity. If Burning Man were to be a hundred-year movement, it had to be bigger than any of its personalities.

A big piece in the puzzle, though, would henceforth be conspicuously missing. Larry Harvey would no longer be choosing the annual theme. That torch had already passed soon after his death to his colleagues in the Philosophical Center. "The right people are in place to continue the particular brilliance Larry had," predicts Will Roger, who stepped down as chairman earlier in the year. "They spent enough time with Larry to understand the dynamics that he used."

The responsibility is now Stuart Mangrum's. Over the years, he and Harvey had perfected a process to come up with each year's theme, always starting with a joke. Now, as he settled in to ponder future themes, Mangrum felt a void. "A joke's only funny if somebody laughs and there's nobody to laugh with me. I keep thinking of things I want to tell him."

The Project will carry on. As far as taking care of business, of mounting the event, the transition has been seamless. The do-ocracy grows more adept. The Renwick show was a big stepping stone into the future. Regionals abound. Burners Without Borders is full of vitality. Black Rock Labs will soon come online. Fly Ranch awaits. Perhaps someday there will be a Burner cryptocurrency or an interactive digital platform or a full-fledged consultancy à la McKinsey to facilitate civic engagement.

Even in Larry Harvey's fertile imagination, none of this could have been conceivable in 1986. What's happened since the bonfire on the beach hasn't been his doing alone. So, so many others contribute to the grand stew. Still, without him, Burning Man would have never existed.

Now he was gone.

"I think that he loved humanity," Will Roger told me when I asked about Harvey's most defining characteristics. "He was always interested in the people he came upon. It didn't matter their walk of life, where they were from, if they were crazy or sane. He had that rare quality of being present." This acceptance was coupled with keen curiosity. "He was open to new ideas and integrating them into his philosophy. That made him unique in the way he looked at the world."

In the Native American tradition, somebody who dies remains alive as long as they're remembered by at least one person. By that measure, Larry Harvey very much lives on. Thousands of people carry at their core a little piece of him. I asked Harley Dubois, as a final postscript, what she would write as an epitaph had he been buried beneath a gravestone. She tried to brush the question aside. Every good reporter knows that in such situations the trick is to stay silent and outwait a reluctant source. Finally, she answered: "A man who sees and feels deeply."

Epilogue

O for a muse of fire that would ascend the brightest
heaven of invention.

WILLIAM SHAKESPEARE,

Henry V

If Larry Harvey had arranged his own memorial service in advance, which his guarded personal privacy and genuine modesty could never have allowed, the heartfelt sincerity of the celebration of his life that unfolded at San Francisco's ornate Castro Theater on the night of July 14 would have exceeded even *his* boundless imagination.

Three days before the event—"The Man with the Hat" was

its name—Marian Goodell reached out to me. Was I coming? I told her I hadn't known anything about it, last-minute travel from D.C. would be difficult, there were family obligations, etc., etc. I'd have to pass. "I understand," she emailed back. "I know there is a slim chance. So I am nudging you!" Nudge taken. Friday morning, I got on a plane.

Saturday night, I was meandering through the Castro District, the city's gay village, where a gigantic rainbow flag waves, waiting for the doors of the theater to open. Like most things Burning Man, the genesis of this occasion was grassroots; two staff alumni, Rosalie Barnes and Will Chase, decided on their own to produce the event. Chase was formerly Burning Man's minister of propaganda. Harvey would often wander over to Chase's desk unannounced and "strike up an impromptu conversation, blowing a good chunk of my day out of the water—leaving in his wake innumerable stories I'll long hold dear." Such tales needed to be heard. So Will and Rosalie organized the tribute as an evening of storytelling.

It began with an entrance processional, led by bagpipe-playing Andrew Johnstone, better known as Haggis (as in the national Scottish dish of oatmeal pudding and sheep's pluck). He's in charge of constructing the Man. "Larry came back every year always wanting something bigger and more audacious." His crew was at work on I, Robot as he spoke, albeit a considerably more modest version than Harvey imagined. "The Man of Harvey's dream," Haggis called it. "He wanted it agile enough to do yoga . . . which would have been possible if we had another million dollars and another year."

Harley Dubois took the stage, offering her take on the infamous 1996 debacle of death and destruction. More than once that week, she recounted, she had wanted to go home. The last night, surrounded by madness, she crawled into her tent sobbing hysterically. Out of nowhere, Harvey appeared and asked if he could enter. "He just hugs me, he holds me as I cry." Eventually she regained composure. And then furiously vented all her anger on him. "I shrieked, I screamed, I said he was responsible. I even insisted that he should write us all thank-you notes."

Hearing herself say this now, she laughed at the ridiculousness of the remark. "Yeah, right. You can imagine Larry ever writing a thank-you note!" She paused. Choking up. Then, in an aside that sounded like a lament, added, "I never ever found out what he thought about that night. He never mentioned it."

Stewart Harvey told family stories. How their father was called "Shorty" because that was the name of the lunch café he owned in Nebraska until it went bankrupt, after which he rigged together a wooden trailer to travel around the next ten years with Harvey's mother as part of the Depression migration until finally settling outside Portland, where the house he built was wiped out within a year by the flooding Columbia River. "Larry had two distinct natures, a happy kid and a performance personality." He spoke of his brother's strong character: "He always valued principles." And of his gift for turning unlikely possibilities into reality: "He convinced creative people that they could do something together that was greater than any of them could do on their own."

The artist Kate Raudenbush recollected how, as a young student excited by Impressionism, she wondered how such a paradigm-shattering art movement happens in the first place. She solved the riddle at Burning Man. "It was doing Big Art when nobody else would." The answer was that there needed to be at the center somebody like Larry Harvey. "He never made Burning Man about him." Instead, he convinced playa artists that the best work "is made in dialogue with ideas." That "creativity connects us, it changes consciousness."

Others took their turns. Jerry James, Harvey's carpenter partner in constructing the beach Burns, confessed he really couldn't explain what prompted Larry to build that first effigy. Jennifer Raiser, treasurer and board member, recounted Larry's delight in going incognito to play a souk merchant giving away the store at the exotic Silk Road bazaar pavilion of 2014. "He was the caretaker," said Jimmy Mason, a longtime friend called to the stage. "He had the best ideas any of us ever knew. He created the best experiences any of us ever knew. He had a higher responsibility to those things."

Last to speak was Tristan, Larry's son. When Larry Harvey and I first met, an arranged meeting in a café up near Columbia University in Manhattan, his off-putting style froze me up. He didn't know me from Adam and here I was proposing to write a book about him and Burning Man. I was getting bad vibes and feared my project was about to implode. Then came an exchange that got us over the hump. We discovered each of us had a child the same age whom we deeply loved, my daughter Katherine and his son. That was the moment we clicked, when we established the kinship of sentiment that paved the way for what became a rich intellectual relationship. This was the side I suspect few but his closest intimates ever saw, Harvey the father. "Tristan was my brother's first great commitment," said Stewart. "Fatherhood may not have come naturally to him, but the desire to succeed as a father figure ran deep."

"I'm gonna give you the truth," Tristan began, voice quaking.

His grief remained palpable. "No Santa, no fairy tales. The way my father did." First, though, he needed to add a footnote to Harley's story. Turns out that he and his father had actually discussed the 1996 incident just a few weeks before the stroke. "He didn't think you were ever coming back." Larry drew a lesson from comforting her, a moral he shared with his son years later: "The people who are important to you, hold them close."

Tristan described the final days, when the doctors needed authority to perform brain surgery. "I was making decisions for my father, I needed affirmation that he agreed with me." So Tristan spoke to him, explaining the situation. Harvey's condition would quickly deteriorate but Tristan felt they were still communicating. Larry raised his left hand and Tristan believed it was a signal giving permission to proceed. "That was our last contact as two independent people."

But in the intervening months, Tristan told the crowd who were holding on to his every word, he had been rethinking that final interaction. He now understood the meaning of that moment differently. Larry was still aware of his surroundings, aware of who was in the room. And knew he was dying. Raising his hand wasn't a response to Tristan's question but, rather, a gesture of farewell. "He was saying goodbye to the people he loved."

ACKNOWLEDGMENTS

As befits its subject, the writing of *Radical Ritual* has been a great adventure. Along the trail, I've been helped by an extraordinary crowd. You know who you are. I apologize in advance for omitting names that should be included; the oversight's due to an abundance of debts rather than an absence of gratitude.

Rosie Lila got the whole thing started that fateful night at First Camp. Marian Goodell was enthusiastic from the get-go; proceeding would have been impossible without her support. Ditto with Megan Miller. Harley K. Dubois, as is her wont, generously showed up when most needed. Stuart Mangrum kept me steering a steady course. Ray Allen greeted each new installment of the work-in-progress with unwavering confidence that exceeded my own. Danger and Dusty went from being sources to friends.

In no particular order, I thank Charles Planck, Matt Gordon, John Hartman, Robert Haferd, and the sisters of 3SP (Lesley Stein, Kathy Baird, and Marie Blakey). Kenny Reff and Vanessa Franking, shooters of the first-rank, contributed wonderful photographs. John Carmen offered much appreciated hospitality whenever I needed to be in San Francisco. Grace and Ron Nichols graciously hosted me in Reno.

Marty Baron's skepticism about the project—"I wish I could understand why you're writing about Burning Man"—inspired me to dig deep to produce good answers. Ralph Eubanks deftly nurtured the manuscript with his keen editorial touch and, as my agent, found it the perfect home. Author-friendly publishers may be an endangered species, but Jack Shoemaker and his colleagues at Counterpoint Press continue to set a high standard.

Having a writer for your father can be challenging; Katherine, Amelia, and William bear the burden well. Being the wife of a writer can be even tougher. Nobody does it better than Cait.

NOTES

I: In the Beginning

1. Brian Doherty, *This Is Burning Man: The Rise of a New American Underground* (BenBella Books, 2006), p. 23.

2. Quoted in "Larry Harvey," www.zpub.com/burn/burn-lh.html.

3. Doherty, *This Is Burning Man*, p. 19–21.

4. Joseph Campbell, *The Hero with a Thousand Faces* (Princeton University Press, 1968), p. 30.

5. Mark Beers, "Who the Hell Is John Law?," *Comet Magazine*, www.cometmagazine.org/cometsite4/cometsite3/comet2/jlaw.html.

6. Gary Warne, "Carnival Cosmology," *The Burning Man Journal*, journal.burningman.org/2015/09/philosophical-center/tenprinciples/carnival-cosmology-by-gary-warne.

7. Gary Warne, "1977 Evolution into Chaos: A Chronology," www

.suicideclub.com/9/essays/1977-evolution-into-chaos-a-chronology
-gary-warne.

8. "Duchamp's urinal tops art survey," BBC, December 1, 2004, news.bbc.co.uk/2/hi/entertainment/4059997.stm.

9. Adam Jacques, "Larry Harvey: The founder of the Burning Man festival on adoption, uncontrollable rage—and how Freud became a father figure," *The Independent*, January 25, 2014, www .independent.co.uk/news/people/profiles/larry-harvey-the-founder -of-the-burning-man-festival-on-adoption-uncontrollable-rage -and-how-freud-9083511.html.

10. Ibid.

11. Brad Wieners, "Hot Mess," *Outside Magazine* Online, August 24, 2012, www.outsideonline.com/1925281/hot-mess.

12. Ibid.

13. Doherty, *This Is Burning Man*, p. 31.

14. Wieners, "Hot Mess."

15. Doherty, *This Is Burning Man*, p. 42.

16. Louis Brill, "The First Year in the Desert," www.burningman.org /culture/history/brc-history/event-archives/1986-1991/firstyears.

17. "Burning Man Burnout," *Wired*, July 7, 1997, www.wired.com /1997/07/burning-man-burnout.

18. Doherty, p. 86; *Black Rock Gazette*, August 29, 1996.

19. Tony Perez-Banuet, *Coyote Nose, Tales of the Early Desert Carnies of Burning Man*, Chapter 4: Inferno, journal.burningman.org /coyotenose.

20. Perez-Banuet, *Coyote Notes*.

21. Doherty, *This Is Burning Man*, p. 98.

22. Perez-Banuet, *Coyote Notes*, Chapter 5: God's Flashlight.

23. Doherty, *This Is Burning Man*, p. 122.

24. Ibid., p. 119.

25. Steven T. Jones, *The Tribes of Burning Man: How an Experimental City in the Desert Is Shaping the New American Counterculture* (CCC, 2011), p. 49.

26. "Burning Man Burnout," *Wired*.

27. Rod Garrett, "Designing Black Rock City," journal.burningman

.org/2010/04/black-rock-city/building-brc/designing-black-rock -city.

28. Fred Bernstein, "A Vision of How People Should Live, from Desert Revelers to Urbanites," *New York Times*, August 28, 2011.

29. www.youtube.com/watch?v=U5P9Bin4JLc.

30. "Burning Man Grows Up," *Business 2.0*, July 1, 2007, money .cnn.com/magazines/business2/business2_archive/2007/07/01 /100117064/.

31. Jones, *The Tribes of Burning Man*, p. 92.

32. Ibid., p. 86.

II: Chaos and Control

1. Lee Gilmore and Mark Van Proyen, *AfterBurn: Reflections on Burning Man* (University of New Mexico Press, 2005).

2. Erik Davis, "Beyond Belief," in *AfterBurn*, p. 26.

3. "How Burning Man Is Going to Destroy Your Relationship," *BRC Weekly: Black Rock City's Independent Newsweekly*, August 29–September 4, 2016.

4. Ibid.

5. "Founded on Fire Magick," www.burners.me/2015/06/24 /founded-on-fire-magick.

6. "David Best, The Man Who Builds Art and Burns It," *The Guardian*, February 14, 2015, www.theguardian.com/artanddesign /2015/feb/15/david-best-the-man-who-builds-art-and-burns-it -burning-man-derry.

7. Ibid.

8. Chip Conley, "Why I Helped Buy Fly Ranch," *Fest 300* magazine, July 21, 2016, www.fest300.com/magazine/why-i-helped -burning-man-buy-fly-ranch.

9. The 10 Principles of Burning Man:

Radical Inclusion

Anyone may be a part of Burning Man. We welcome and respect the stranger. No prerequisites exist for participation in our community.

Gifting

Burning Man is devoted to acts of gift giving. The value of a gift is unconditional. Gifting does not contemplate a return or an exchange for something of equal value.

De-commodification

In order to preserve the spirit of gifting, our community seeks to create social environments that are unmediated by commercial sponsorships, transactions, or advertising. We stand ready to protect our culture from such exploitation. We resist the substitution of consumption for participatory experience.

Radical Self-reliance

Burning Man encourages the individual to discover, exercise, and rely on his or her inner resources.

Radical Self-expression

Radical self-expression arises from the unique gifts of the individual. No one other than the individual or a collaborating group can determine its content. It is offered as a gift to others. In this spirit, the giver should respect the rights and liberties of the recipient.

Communal Effort

Our community values creative cooperation and collaboration. We strive to produce, promote, and protect social networks, public spaces, works of art, and methods of communication that support such interaction.

Civic Responsibility

We value civil society. Community members who organize events should assume responsibility for public welfare and endeavor to communicate civic responsibilities to participants. They must also assume responsibility for conducting events in accordance with local, state, and federal laws.

Leaving No Trace

Our community respects the environment. We are committed to leaving no physical trace of our activities wherever we gather. We clean up after ourselves and endeavor, whenever possible, to leave such places in a better state than when we found them.

Participation

Our community is committed to a radically participatory ethic. We believe that transformative change, whether in the individual or in society, can occur only through the medium of deeply personal participation. We achieve being through doing. Everyone is invited to work. Everyone is invited to play. We make the world real through actions that open the heart.

Immediacy

Immediate experience is, in many ways, the most important touchstone of value in our culture. We seek to overcome barriers that stand between us and a recognition of our inner selves, the reality of those around us, participation in society, and contact with a natural world exceeding human powers. No idea can substitute for this experience.

10. Lewis Hyde, *The Gift: Imagination and the Erotic Life of Property* (Random House, 1979).

11. "Burning Man Turns 30: The Joys, Pitfalls (and Drugs) of Hollywood's 'Vacation for the Soul,'" *Hollywood Reporter*, August 16, 2016, www.hollywoodreporter.com/features/burning -man-turns-30-joys-922027.

12. Katherine K. Chen, *Enabling Creative Chaos: The Organization Behind the Burning Man Event* (University of Chicago Press, 2009).

13. Rakesh Khurana, review quote on back cover of Chen, *Enabling Creative Chaos*.

14. John Markoff and G. Pascal Zachary, "In Searching the Web, Google Finds Riches," *New York Times*, April 13, 2003, www .nytimes.com/2003/04/13/business/in-searching-the-web-google -finds-riches.html.

15. Fred Turner, "Burning Man at Google: A Cultural Infrastructure for New Media Production," *New Media and Society*, 2009, fredturner.stanford.edu/wp-content/uploads/turner-nms-burning-man.pdf.

16. Ibid.

17. Nicole Laporte, "The Woman Behind the Superlatives: Three Things You Need to Know About Susan Wojcicki," *Fast Company*, August 6, 2014, www.fastcompany.com/3033957/the-woman-behind-the-superlatives-three-things-you-need-to-know-about-susan-wojcicki.

18. Laszlo Bock, *Work Rules! Insights from Inside Google That Will Transform How You Live and Lead* (Twelve, 2015), p. 28.

19. Jay Yarrow, "Google CEO Larry Page Wants a Totally Separate World Where Tech Companies Can Conduct Experiments on People," *Business Insider,* May 16, 2013, www.businessinsider.com/google-ceo-larry-page-wants-a-place-for-experiments-2013-5.

20. Turner, "Burning Man at Google."

III: Burning Man's Utopian Vision

1. Doherty, *This Is Burning Man*, p. 24.

2. "Black Rock City Honoraria," www.burningman.org/culture/burning-man-arts/grants/brc-honoraria.

3. "That's Show Business," *The Guardian*, June 30, 2004, www.theguardian.com/artanddesign/2004/jun/30/art1.

4. "Burners Without Borders—Dispatches from Peru," *The Burning Man Journal*, journal.burningman.org/2010/10/global-network/burners-without-borders/burners-without-borders-dispatches-from-peru.

5. Leah Meisterlin, "Anti-Public Urbanism: Las Vegas and the Downtown Project," *The Avery Review*, www.averyreview.com/issues/3/antipublic-urbanism.

6. Peter Hall, "IDEO, Making Government More Innovative, Less Bureaucratic," *Metropolis*, www.metropolismag.com/ideas/ideo-takes-on-the-government.

7. "How Burning Man Spawned a Solar Gold Rush," www
.nationswell.com/burning-man-spawned-solar-gold-rush.

8. "On Future of Energy, from U.S. 'Renaissance' to the Rise of Renewables," *New York Times*, November 14, 2017.

9. Jack Zenger and Joseph Folkman, "Do Women Make Bolder Leaders Than Men?," *Harvard Business Review*, April 27, 2016, www.hbr.org/2016/04/do-women-make-bolder-leaders-than -men.

10. Craig Rullman, "High and Lonesome," *Bunkhouse Chronicle*, August 19, 2014, www.thebunkhousechronicle.com/2014/08/19 /high-and-lonesome.

11. Rullman, "High and Lonesome."

12. Amy Henderson, "No Spectators for Smithsonian American Art Museum's Burning Man," *Artes*, April 6, 2018.

13. Michael O'Sullivan, "This Exhibit Brings the Spirit of Burning Man to D.C. Well, Minus the Drugs, Sex and Desert," *The Washington Post*, March 29, 2018.

14. Rachel Tashjian, "A Guide to What You're Missing Out on at Art Basel Miami Beach," *Vanity Fair*, December 1, 2014.

15. Grover Norquist, "My First Burning Man: Confessions of a Conservative from Washington," *The Guardian*, September 2, 2014, www.theguardian.com/commentisfree/2014/sep/02/my-first -burning-man-grover-norquist.

16. Jon Bond, "An ad executive explains how Burning Man is really a massive marketing festival," *Business Insider*, September 6, 2017, www.businessinsider.com/an-ad-exec-explains-how -burning-man-is-really-a-marketing-festival-2017-9.

17. Daniel Terdiman, "How a Burning Man Camp Project Became a Multimillion Dollar Business," *Fast Company*, August 22, 2017, www.fastcompany.com/40448192/how-a-burning-man -camp-project-became-a-multimillion-dollar-business.

18. Stephanie Malik, "Manish Arora on How Burning Man Changed His Life & Inspired His Latest Collection," *Bullett*, March 5, 2013, www.bullettmedia.com/article/manish-arora

-on-how-burning-man-changed-his-life-inspired-his-latest
-collection.

19. Daniel Pinchbeck, "Why I Consider Burning Man the Greatest Cultural Movement of Our Time," October 11, 2016, www
.artsy.net/article/artsy-editorial-why-i-consider-burning-man
-the-greatest-cultural-movement-of-our-time.

INDEX

The Seattle Public Library
Beacon Hill Branch
Visit us on the Web: www.spl.org

Check out date: 08/29/19

xxxxxxxxx4443

Upheaval : turning points for nation
0010095308143   Due date: 09/19/19
book

Radical ritual : how Burning Man cha
0010099881756   Due date: 09/19/19
book

TOTAL ITEMS: 2

Renewals: 206-386-4190
TeleCirc: 206-386-9015 / 24 hours a day
Online: myaccount.spl.org

* * * * * * * * * * * * * * * * *

Pay your fines/fees online at pay.spl.org

NEIL SHISTER has been a correspondent with *Time*, television critic for the *Miami Herald*, and editor of *Atlanta* magazine. He's taught at Hampshire College, Boston University, and The George Washington University. He was a Peace Corps volunteer. He lives in Washington, D.C., with his wife and son.